STUDENT'S SOLUTIONS MANUAL

DAVID LUND

University of Wisconsin – Eau Claire

STATISTICAL REASONING FOR EVERYDAY LIFE

FOURTH EDITION

Jeffrey O. Bennett

University of Colorado at Boulder

William Briggs

University of Colorado at Denver

Mario F. Triola

Dutchess Community College

PEARSON

Boston Columbus Indianapolis New York San Francisco Upper Saddle River
Amsterdam Cape Town Dubai London Madrid Milan Munich Paris Montreal Toronto
Delhi Mexico City São Paulo Sydney Hong Kong Seoul Singapore Taipei Tokyo

www.pearsonhighered.com

PEARSON

PREFACE

Student's Solutions Manual to Accompany Statistical Reasoning for Everyday Life by Jeffrey Bennett, William Briggs, and Mario Triola provides solutions to every odd-numbered exercise in Chapters One through Ten of the text and to the odd-numbered Review Exercises and Quiz Problems at the end of each chapter. The solutions are more than just answers -- reasoning and intermediate steps in the process of solving the exercises are also presented.

I want to express my thanks to my wife, Judy, for her support and patience throughout this project, and to Ely Delman and Joe Vetere at Pearson Education who have been such a great help in bringing this work to completion.

CONTENTS

CHAPTER 1 ANSWERS

<u>Section 1.1</u>
Statistical Literacy and Critical Thinking

1 A population is the complete set of people or things being studied, while a sample is a subset of the population. The difference is that the sample is only a part of the complete population.

3 A *sample statistic* is a characteristic of a *sample* found by consolidating or summarizing raw data. A *population parameter* is a characteristic of an entire *population*. Since it is usually impractical to obtain raw data for entire large populations, it is also not likely that population parameters can be directly measured. For that reason, we use measured sample statistics to make inferences about the values of population parameters.

5 This statement does not make sense. Population parameters are inferred from sample statistics, so it's not possible to have the former without the latter. The only way to determine a population parameter is to obtain raw data for every individual in the population, in which case there is no error at all.

7 This statement does not make sense. The poll makes it seem like Johnson should win the election because the confidence interval for the percent of voters voting for Johnson runs from 54% - 3% to 54% + 3% (51% to 57%), suggesting that she should have obtained more than half of the votes, enough to win. However, in most cases such as this, the margin of error is defined to mean that we can be 95% confident that the true percent of votes lies in the range from 51% to 57%. Because 5% of the time a 95% confidence interval will not contain the actual percent of votes, that percent could be above 57% or below 51%. If, in fact, it does lie below 51%, it could also be below 50%, in which case Johnson loses the election.

9 This statement does not make sense. The population of interest is all people who have suffered a family tragedy, not only people who have suffered the loss of a spouse and are in a support group. There are other types of family tragedies besides the loss of a spouse, and not all of the people suffering those tragedies join support groups. The sample must be taken from the population of interest

Concepts and Applications

11 The sample consists of the 1018 adults in the U.S. who were surveyed. The population consists of all adults in the U.S. The sample statistic is the 22% who said that they had smoked in the past week. The value of the population parameter is not known, but it is the percentage of all adults in the U.S who smoked in the past week.

13 The sample consists of the 47 subjects treated with Garlicin. The population consists of all adults. The sample statistic is the 3.2 mg/dL mean drop in the level of LDL. The population parameter is unknown, but it is the mean drop in the level of LDL.

15 The range of values likely to contain the true value of the population parameter is from 60% - 3% to 60% + 3% or from 57% to 63%.

17 The range of values likely to contain the true value of the population parameter is from 96% − 3% to 96% + 3% or from 93% to 99%.

19 Yes. Although there is no guarantee, the results suggest that the majority of adults believe that immediate government action is required, because the interval from 53% to 57% most likely contains the true percentage.

21 With a sample statistic of 70% and a margin of error of 3 percentage points, we are 95% confident that the interval from 67% to 73% contains the

Copyright © 2014 Pearson Education, Inc.   **1**

population parameter that is the true percentage of the voters who would say that they voted in the recent presidential election. This entire range is, however, somewhat higher than the actual 61% who voted according to the voting records. This suggests that there were some people in the sample who did not actually vote, but said that they did when polled. While it is still possible (as always) that this particular sample is unusual and everyone told the truth, the lower end of the range (67%) is quite far from 61%, making this an unlikely possibility.

23   a)   The goal was to determine the percentage of all adults in favor of the death penalty for people convicted of murder. The population is the complete set of all adults and the population parameter is the percent of those adults in favor of the death penalty for people convicted of murder.

    b)   The sample consists of the 511 selected adults. The raw data consists of those subjects' responses to the question and the sample statistic is the 64%.

    c)   The range of values likely to contain the population parameter is from 64% − 4% to 64% + 4% (or from 60% to 68%).

25   a)   The goal is to determine the percentage of households with a TV tuned to the Super Bowl game. The population consists of the set of all U.S. households, and the population parameter is the percentage of those households with a TV tuned to the Super Bowl game.

    b)   The sample consists of the 9,000 households surveyed. The raw data consist of the individual indications of whether or not the individual household has a TV tuned to the Super Bowl game. The sample statistic is the percentage of households in the sample with a TV tuned to the game, 45%.

    c)   The range of values likely to contain the population parameter is 45% $\pm$ 1% or 44% to 46%.

27   Great care must be used in designing a valid survey. The time and location of the survey may be critical factors that influence the results. Obviously, no one will be using a cell phone in an area where there is no service. Few drivers will use them while driving in the middle of the night. Many may be using them while caught in a rush hour traffic jam. There may be other factors that can influence the results, but these are a few examples.

    Step 1:   Goal: Determine the percentage of all drivers who use cell phones while they are driving.

    Step 2:   Choose a sample of drivers while they are driving.

    Step 3:   Somehow, observe the drivers in the sample to determine whether or not they are using a cell phone at the time of the observation. Note that this may be difficult if the driver is using a hands-off device.

    Step 4:   Use statistical techniques to infer the likely percentage of all drivers who are using cell phones while driving.

    Step 5:   Based on the likely value for the population parameter, draw conclusions about the percentage of drivers who use cell phones while they are driving.

29   Step 1:   Goal: Determine the mean weight of airline passengers.

    Step 2:   Choose a sample of airline passengers.

    Step 3:   Weigh each selected passenger and calculate the mean weight of those in the sample.

    Step 4:   Use statistical techniques to make inferences about the mean weight for the entire population of airline passengers.

    Step 5:   Based on the likely value of the population mean, form a conclusion about the average weight of all airline passengers.

Section 1.2
Statistical Literacy and Critical Thinking

1 A *census* is the collection of data from every member of the population. A sample is the collection of data from some, but not all, members of the population. For a given population, a sample will contain less data than will a census.

3 Cluster sampling involves randomly selecting subgroups of a population and then selecting all members of the population in each subgroup. For example, one might randomly select some city blocks and then interview all people living on those blocks. Stratified sampling involves randomly selecting members from each of different subgroups (or strata) of the population. For example, one could randomly select some men and randomly select some women, keeping the results separate for each of the two gender subgroups.

5 This statement does not make sense. A census would mean getting age data for every person who earns a bachelor's degree in the country (or world), which is clearly not practical or possible.

7 This statement makes sense. It's quite apparent that most Americans are not more than 6 feet tall, so a study that comes to a ridiculous conclusion must have suffered from some form of bias.

Concepts and Applications

9 Since the number of players on the LA Lakers is small, a census is practical, and it is easy to obtain their heights (for example, from a Laker website).

11 A census is not practical since it would require obtaining the IQ of every statistics instructor in the U.S. The number of statistics instructors is very large and it would be difficult to get them all to take an IQ test.

13 The sample consists of the service times of the four selected Senators. The population consists of the service times of all 100 Senators. The sampling method is simple random sampling. Since the sample is so small, there is a good chance that it is not representative of the entire population.

15 The sample consists of the 1059 randomly selected adults. The population consists of all adults. Simple random sampling was used. Because the sample size is quite large and sampling was done by a well-established and reputable firm, the sample is likely to be representative of the population.

17 The most representative sample is likely to be Sample 3 because the list will contain people from all over Florida and there is no reason to suspect that the people with the first 1000 numbers would differ in any particular way from the other people. [This assumes that the list is alphabetical, not in order by phone number, in which case the first three digits of the phone are likely to be the same and the entire sample would come from one area, possibly one city, of Florida.] Sample 1 is biased because it involves owners of expensive vehicles. Such owners may be able to pay off their credit cards monthly or may be people with greater credit limits on their credit cards. Sample 2 is biased because it includes only people from the Fort Lauderdale area. Sample 4 is biased because it includes only people who are self-selected and may have strong feelings about the issue of credit card debt.

19 The critic may be under real or imagined pressure to give a favorable review to the film since she works for the same company that produced the film.

21 The university scientists are receiving funding from Monsanto, which might make them eager to please Monsanto in hopes of getting additional funding opportunities in the future. Thus, there is a potential for bias toward giving Monsanto the results it wants, even though they do not work for Monsanto.

23   This sample is a simple random sample that is likely to be representative because there is no inherent bias in the selection process.

25   This is an example of cluster sampling. It is likely to be unbiased as long as there are enough polling stations selected for the sample so that the entire sample has a chance to be representative on a national level. Since the actual results of the election are usually known within a few hours of exit poll results and the exit polls are unlikely to influence the voting of any other voters, poor sampling techniques that have a good chance of resulting in embarrassment for the news media are likely to be avoided.

27   This is an example of convenience sampling. It is likely to be biased because members of a family are likely to be more similar in their physical characteristics and strength than would a sample taken from the population as a whole.

29   This is a stratified sample with the strata being the various age groups. As described, the sample is likely to be biased because it contains equal numbers of people in each of the age groups whereas the population is not equally distributed among these age groups. There are two ways to remedy this problem. The results from the age strata could be combined by "weighting" the results from each stratum to reflect the sizes of the strata in the population as a whole. A second way is use proportionate sampling in which each stratum in the sample has a number of members that is proportional to its presence in the population as a whole.

31   This sample is a systematic sample. It is unlikely to be biased because there is nothing about an alphabetical list that is likely to produce a biased sample when testing a telemarketing technique.

33   This is a stratified sample. It is likely to be a biased sample because population does not consist of employed, unemployed, and employed part time in equal numbers. It is possible to correct this bias by "weighting" the strata results to reflect the strata sizes in the population.

35   This is a convenience sample, and it is one that is likely to be biased because people with strong feelings are more likely to return the survey. The magazine probably chose this sampling method because it was easy; the magazine might even be interested in the opinions of those with the strongest feelings.

37   This is a simple random sample and is therefore likely to be representative. The sample size is not specified, but the larger the sample size, the better the chance that the sample is representative.

39   Simple random sampling should be adequate for a student election if the sample is large enough.

41   Since all states have single departments that keep all death records, it should be easy to randomly select some states and then search the computer records to determine the number and percentage of deaths due to heart disease each year. This is an example of cluster sampling with each cluster being a state. [The U.S. Center for Disease Control (CDC) routinely collects these data from all states and they are available on the CDC website.]

Section 1.3
Statistical Literacy and Critical Thinking

1   A placebo is physically similar to a treatment, but it lacks any active ingredient, so it should not have any effects on the subject. A placebo is important so that results from subjects given a real treatment can be compared to the results from subjects given a placebo.

3   Confounding occurs when it is not possible to ascertain what caused the effects that were observed. In this instance, if males were chosen for the real treatment and females were chosen for the placebo group, and if a difference resulted in the effects on the two groups, it would not be possible to tell whether those effects were caused by the treatment or by the gender of the subjects.

5   It almost always makes sense to use double blinding for an experiment, but it is sometimes impossible or difficult to do. In this case, both subjects and experimenters can see the clothing worn by the subjects. Blinding must therefore be achieved by some other method. The subjects may be blinded by not telling them the purpose of the experiment or even that there is an experiment so that their knowledge of the color of their clothes does not affect the results. The experimenters clearly know the purpose of the experiment, so blinding is not possible for them. It is therefore necessary that data be based on objective measures that are not influenced by any judgments of the experimenters.

7   The experimenter effect occurs when the psychologist somehow influences subjects by such things as tone of voice, facial expressions, or attitude. It can be avoided by using blinding so that those who evaluate the results do not know which subjects are given an actual treatment and which subjects are given no treatment or a placebo. It might also help if the subjects responded to written, rather than oral, questioning or to a computerized voice that conveys exactly the same attitude to every subject and does not have different tones of voice or facial expressions associated with it.

Concepts and Applications

9   This is an observational study because the batteries were tested, but they were not given any treatment.

11   This is an experiment. There is a treatment group of subjects that received the magnetic bracelets and a placebo group that received the non-magnetic bracelets. The variable of interest is whether or not the passengers experienced motion sickness. Blinding might not be totally successful since passengers might happen to detect whether their bracelets are magnetic by holding them near something made of iron.

13   This is an observational, retrospective study examining how a characteristic determined before birth (fraternal or identical twins) affected mental skills later.

15   This is an experiment because the subjects were given a treatment. The experimental group consists of the 152 couples who were given the YSORT treatment, and the control group consists of others not given any treatment.

17   This is an experiment. The treatment group consists of the Bt corn and the control group consists of corn not genetically modified.

19   This is an experiment since the subjects received different treatments. The treatment group consists of the individuals given the magnetic devices and the control group consists of those given the non-magnetic devices.

21   Confounding is likely to occur. If there are differences in tree growth in the two groups, it will be impossible to tell if those differences are due to the treatment (fertilizer or irrigation) or to the type of region (moist or dry). This confounding can be avoided by using blocks of fertilized trees in both regions and blocks of irrigated trees in both regions.

23   Confounding is likely. If there is a difference in the amount of gasoline consumed between the two groups, it will not be possible to tell whether the difference is due to the type of vehicles in the two groups or to the octane rating of the gasoline used. Confounding can be avoided by using 87 octane gasoline in half of the vehicles in each group and 91 octane gasoline in the other half. It would be even better to have all individual vehicles driven under identical conditions, once with the 87 octane gasoline and once with the 91 octane gasoline.

25   Subjects clearly know whether they are treated with running, so confounding

is possible from a placebo effect. Moreover, there is no objective way to measure back pain, so different subjects may report changes in back pain differently. Also, there could be an experimenter effect that can be avoided with blinding of those who evaluate results.

27   Confounding is possible. If a difference is found in the effects on blood pressure from lifting weights or tennis balls, you want to ensure that the difference is a result of the two treatments, not from some subjects' apprehension over having their blood pressure measured or from an experimenter's judgment of the effect on blood pressure. The experimenter effect can be avoided by using technology and trained personnel to measure the blood pressure without any interaction with the experimenter. The placebo effect can be reduced or eliminated by having the same subjects use the heavy weights and tennis balls at different times, with the order mixed. Any apprehension over the measurement process should be the same for both sets of weights and will therefore be canceled out.

29   The control group consists of those who do not listen to Beethoven, and the treatment group consists of those who do listen to Beethoven. Blinding of the subjects is automatic since the infants won't know they are part of an experiment. By coding the subjects, blinding could be used so that those who measure intelligence are not influenced by any knowledge of which group the subjects were in. There is an additional problem that could arise in interpreting the data from the experiment. If a difference between the two groups is found, is it a result of listening to Beethoven, or is it a result of just listening to some kind of music? As designed, this experiment will not be able to determine the answer. If the real interest is Beethoven's music, the experiment must be expanded to include more groups with other kinds of music.

31   The control group consists of a group of cars using gasoline without the ethanol additive. The treatment data should be obtained by using the same cars with gasoline containing the ethanol additive, mixing the order in which the two gasoline blends are used for the different cars. In this way, it is possible to ensure that any observed difference in mileage is the result of the difference in gasoline blend. If two different groups of cars were used, a difference in mileage between the two groups might have been caused by differences in the cars themselves, even if they were all of the same brand and model. There is no need to blind the cars, and since the mileage will be determined without any judgments on the part of the experimenters, there is no need to blind the experimenters.

Section 1.4
Statistical Literacy and Critical Thinking

1   Peer review is a process by which experts in a field evaluate a research report before the report is published. It is useful for lending credibility to the research because it implies that other experts agree that it was carried out properly.

3   When participants select themselves for a survey, those with strong opinions about the topic being surveyed are more likely to participate, and this group typically is not representative of the general population.

5   This answer does not make sense. A survey involving a large sample could be poor if it involves a poor sampling method such as convenience sampling or a self-selected sample. A smaller sample might yield much better results if it involves a sound sampling method such as simple random sampling.

7   This does not make sense. Often we don't even know if there are confounding variables, let alone how many, so we can't know for certain that they have all been taken into account.

9   The survey was funded by a source that can benefit through increased sales fostered by the survey results, so there is a potential for bias in the survey. Thus Guideline 2 (Consider the source..) is the most relevant.

11 Guideline 4 is most relevant since "good" is not well defined and is difficult to measure. Guideline 4 is most relevant.

13 Guideline 3 is the most relevant since the sample is self-selected, resulting in possible participation bias.

15 Guideline 6 is the most relevant since the wording of the question is biased and intended to elicit negative responses.

17 Because companies involved in the chocolate business provided much of the funding for the research, the researchers may have been more inclined to provide favorable results, to search for only positive aspects of eating chocolate, or to report only results that would be deemed positive by the companies. The bias could have been avoided if the researchers were not paid by the chocolate manufacturers. If that was the only way to fund the research, then the researchers should institute procedures to ensure that they submit all results for publication, including any negative ones.

19 The wording of the question was biased to strengthen opposition to a particular candidate, and is likely to be a "push poll" financed by supporters of another candidate, rather than a legitimate poll. A better sampling method would involve questions devoid of such bias.

21 The results are not necessarily contradictory, but might appear to be so. The word "wrong" in the first question could be misleading or confusing. Some people might believe that abortion is wrong, but still favor choice. Such people would respond "yes" to the first question and "no" to the second. The second question could also be confusing, as some people might think that "advice of her doctor" means that the woman's life is in danger, which could alter their opinion about abortion. Groups opposed to abortion would be likely to cite the results of the first question, while groups favoring choice would be more likely to cite the results of the second question.

23 The first question requires a study of Internet dates generally, while the second examines people who are married to see whether their first date was an Internet date. The first question is a more difficult one to study since, at the time of the study, some Internet dates will not yet have led to marriage, but may eventually. In addition, there is no good way to determine who is in the population of Internet daters. Assuming that the population of Internet daters could be identified, the first question would only tell you how often internet dating leads to marriage, which might not be any different from how other forms of dating lead to marriage. The goals of the study need to be better defined, and the questions framed to meet the goals.

25 The first question involves a study of college students in general, and the second question involves a study of those who do binge drinking. The first question might be addressed by surveying college students. The second question would be addressed by surveying binge drinkers, and it would be much more difficult to survey or even identify this group.

27 The headline says "drugs" whereas the story says "drug use, drinking, or smoking." Because "drugs" is usually taken to mean drugs other than smoking or alcohol, the headline is very misleading. Also note that the headline says "98% of movies" while the story says "98% of top movie rentals", a much smaller set of movies.

29 No information is given about what the "confidence" refers to. For example, does it mean that the public is confident about the military leaders only in military situations, or in other situations (such as business or politics) as well? The sample size and margin of error are also missing in the report, but even if they were present, we still don't know what "confidence" refers to.

31   No information is given to justify the statement that "more" companies try to bet on weather forecasting. If only the four cited companies are new, the increase is certainly insignificant.

Chapter 1 Review Exercises

1   a)   The range of values likely to contain the proportion of all adults with tattoos is from 14% - 2% to 14% + 2% or from 12% to 16%.
   b)   The population consists of all adults in the U.S.
   c)   This is an observational study because the subjects were not treated or modified in any way. The variable of interest is whether the subject has a tattoo, which for this study, can take on either of two values, *yes* or *no*.
   d)   The 14% is a sample statistic based on the sample of 2320 adults, not the population of all adults.
   e)   No. In this case, the sample would be self-selected with a likely participant bias.
   f)   A perfect simple random sample of all adults in the U.S. is probably not possible since some have no phones and there are some with no addresses. You not only need a list from which to choose participants, but also a way to contact them. However, the percentage with phones or addresses is very high, so a sample taken from the population of those with phones will likely yield a sample that is very representative of the population. Therefore, we can use a computer to randomly generate telephone numbers, call those numbers until a desired sample size of adults has been contacted. You could also have a computer generate random social security numbers, identify the people with those numbers, and select those people. However, since many children now have social security numbers, any number corresponding to someone not yet an adult could not be included in the sample. Getting the Social Security Administration to part with the identities of those selected may also be a problem.
   g)   We could stratify the sample by state, taking a simple random sample of adults in each state.
   h)   Select all of the adults in each of a number of random selected voting precincts or streets or roads.
   i)   Systematic sampling would be difficult for the entire U.S. since it would require an ordered list of all adults in the U.S. However, one could select every 10th address on each street in a city. Such a sample would be systematic, but it would not be very representative of the population of U.S. adults as a whole.
   j)   Select your classmates. Again, this type of sample will not be representative of all U.S. adults.

3   a)   No. There is no information about the occurrence of headaches among people who do not use Bystolic.
   b)   Because the headache rate is about the same among Bystolic users as among the placebo group, it appears that headaches are not an adverse reaction to Bystolic use.
   c)   This is an experiment because the subjects are given a treatment.
   d)   With blinding, the participants do not know who is receiving Bystolic and who is getting the placebo. This is important so that a placebo effect is minimized. It is important also that those who evaluate the results do not skew their opinion of the results by their knowledge of who received Bystolic or by assigning subjects to the experimental or control group based on their knowledge of the subjects' condition. If this is done, we have double-blinding.

e)   An experimenter effect occurs if the experimenter somehow influences subjects through such factors as facial expression, tone of voice, or attitude. It can be avoided through the use of blinding.

Chapter 1 Quiz

1   (c)   The sample is a subset of the population. Since the 1200 people were drawn from all of the college students in California, the population of interest is the set of all college students in California.

3   (a)   A sample is representative of the population if it yields results similar to those that would be found if the entire population were studied. A large sample does not guarantee this, nor does a sample chosen in the best possible way, but both increase the chance that the sample is representative.

5   (c)   The experiment is not blind since one group is told about the incentive for perfect attendance, while the other group is told that they are part of an experiment, but clearly they will find out that they are the control group.

7   (c)   A placebo is not supposed to have any effect at all. If some people in the control group experience a result that is supposed to happen only in the experimental group, that is called a placebo effect. Placebos are not supposed to cure warts, so if some people in the control group have warts that are cured even though they haven't received a treatment designed to cure warts, then we have a placebo effect.

9   (b)   We could be 95% certain from Poll X that Powell will receive between 46% and 52% of the vote, while we could be 95% certain from Poll Y that she will receive between 50% and 56% of the vote. Both polls will be correct if she receives between 50% and 52% of the vote, so the results of the polls are not inconsistent with one another.

11   (b)   The conclusion may be valid even if the study was biased. Since Exxon Mobil may have a vested interest in the results of the study, there will be a suspicion that the results reflect the company's interests. Answer C doesn't say anything because we don't know what "it" is.

13   (b)   If you are measuring the weights of cars, the variable of interest is the weight of a car.

15   (b)   Whenever we do a statistical study using a sample from a population, there is always a small chance, even when everything is done correctly to try to ensure that the sample is representative of the population, that the conclusions drawn about the population based on the sample results are not correct.

CHAPTER 2 ANSWERS

<u>Section 2.1</u>
Statistical Literacy and Critical Thinking

1   Qualitative data consist of values that can be placed in different non-numerical categories, (such as male, female, or Democrat, Republican), whereas quantitative data consist of values representing counts or measurements.

3   Yes. Data consist of either qualities or quantities (numbers), so all data are either qualitative or quantitative.

Concepts and Applications

5   Blood groups are qualitative because they don't measure or count anything.

7   Reaction times are quantitative because they consist of measurements.

9   The answers to multiple choice questions are qualitative because they don't measure or count anything.

11   The television shows are qualitative data because they don't measure or count anything.

13   Head circumferences are quantitative because they consist of measurements.

15   The grade point averages are quantitative data since they are measures of course grades.

17   These data are discrete because they are obtained by counting the numbers of checked baggage pieces.

19   The numbers of flights are discrete because only the counting numbers (non-negative integers) are used. No values between any of these numbers are possible.

21   These data are continuous because they are measurements of time and can be any value within some range of values.

23   The numerical test scores are discrete data because they can be counting numbers only.

25   The speeds of cars are continuous data because they are obtained by measuring and can be any number within some range of values.

27   The numbers of stars are discrete data because they can be counting numbers only.

29   Weights of textbooks are at the ratio level of measurement because there is a true zero.

31   Types of movies are qualitative data that cannot be ordered and thus can only be at the nominal level of measurement.

33   Classifications of cars as subcompact, compact, intermediate, or full-size are ordinal data because they can be put in an increasing order. They are not at the interval level because there is no way to say the difference between a subcompact and a compact car is the same as the difference between a compact and an intermediate car.

35   Final course grades are at the ordinal level because they are qualitative data that can be arranged in a meaningful order. We can say that an A is higher than a B, but the difference between an A and a B is not necessarily the same as the difference between a B and a C (or even between another A and B).

37   Social Security Numbers are at the nominal level of measurement. Each one represents the name of a person.

39   Numbers of words spoken are at the ratio level of measurement because there is a true zero.

41 The ratio level does not apply. The ratio is not meaningful because the stars don't measure or count anything. Differences between star values are not meaningful, so the ratings are not even at the interval level, let alone the ratio level.

43 The ratio level does not apply. IQ scores do not measure intelligence on the type of scale required for the ratio level. Essentially, they tell you where a person's intelligence falls within the population, so neither differences nor ratios are meaningful.

45 Since the age of anything cannot be less than zero, there is a true zero for age and the ratio level applies to ages.

47 The ratio level applies to salaries since there is a true zero for salaries. A salary of zero means "no pay."

49 Times of runners are quantitative data, are on a ratio level of measurement, and are continuous. (There is a true zero and fractional values of seconds are possible.)

51 The ID numbers are qualitative and are at the nominal level of measurement. They are just another way to express the employees' names.

53 The years in which employees were hired are quantitative and are at the interval level of measurement. The data are discrete because they consist of whole numbers only. The years are measured from an arbitrary reference point for year 0, not a natural zero starting point. Differences between the years are meaningful, but ratios are not meaningful.

55 The rating data are qualitative at the ordinal level of measurement. The ratings can be ordered, but they are not counts or measurements.

Section 2.2
Statistical Literacy and Critical Thinking

1 It is a random error because there is no way to predict when the clerk will write a number incorrectly, or whether the number will be too high or too low.

3 Accuracy is measured by the relative error, $(1.2034278 - 1)/1 = 0.2034278$ or 20.3%. Thus the measurement is off by 20.3%. This is not very accurate given the ease of carrying out the measurement. The measurement is very precise because the recorded height has 7 decimal places.

5 This statement does not make sense. The number is too precise. Someone has found references to this many species, so there should be at least this many. However, there may be some species that are known to only a small number of people, and there may be some species that have not yet been found.

7 This statement is reasonable. If errors are random, we expect that about half of the errors will be in favor of the supermarket and half will be in favor of the customer.

Concepts and Applications

9 Mistakes should lead to random errors. Dishonesty should lead to systematic errors that benefit the taxpayer.

11 Because about half of the batteries have voltages higher than 3.7 volts and half have less than 3.7 volts, these appear to be random errors in measurement or random errors in the manufacturing process.

13 Random errors could occur when reported incomes are recorded incorrectly or when survey respondents don't know their exact donations. Systematic errors could occur when people tend to report higher donations than they actually have in order to lower their taxable income and pay less in taxes.

15 Random errors could occur when passengers don't know their exact weight; systematic errors could result when passengers lie about their weight by reporting a weight that is considerably lower than their actual weight.

17 Random errors could occur with an inaccurate radar gun or with honest mistakes made when the officer records the speeds. Systematic errors could occur with a radar gun that is incorrectly calibrated so that it consistently reads too high or too low.

19   Random errors occur with errors in calculations or by cigarettes that are bought outside the city and smoked inside (or vice versa). A systematic error might occur if there are cigarettes illegally obtained without tax stamps and no tax is therefore reported. The number of packs sold would then be systematically underreported.

21   The absolute error is \$1750. The correct bill is \$2995 - \$1750 = \$1245. Therefore the relative error is \$1750/\$1245 = 1.41 or 141%.

23   The absolute error is \$1.75 - \$2.75 = -\$1.00; the relative error is -\$1.00/\$2.75 or -36.4%.

25   a)   These errors are random. If they were systematic, there would be a tendency for the measurements all to be too high or all to be too low.

   b)   It is better to report the average of the 25 measurements. It is likely to be in error by less than most of the individual measurements and is more reliable than any single measurement.

   c)   Systematic errors might result from a problem with the measuring device or with the definition of the "length of the room" or from measuring along a path that is not perpendicular to the end walls of the room. They could also occur if all the students had been incorrectly taught how to use the tape measure.

   d)   No. If there is a systematic error in the measurements, that same error will be present in the average.

27   The Department of Transportation is more precise because its weights are given to the nearest 0.1 lb. This is more precise than the nearest 10 pounds of the manufacturer's scale. The manufacturer's scale is more accurate because it is in error by only 23 pounds while the DOT scale is in error by 25.2 pounds.

29   The digital scale is more precise (0.01 kg < .5 kg), and the digital scale is more accurate since 52.88 is closer to 52.55 than 53 is to 52.55. (This assumes that your actual weight is what you thought it was.)

31   No one could possibly know the exact population of the United States today, let alone in 1860, so the claim as to the exact value of the population is not believable.

33   The given number is very precise, but it is not likely to be accurate. There are many people in China who are not counted. In addition, the population of China changed often during the year and probably during the census. The census of any nation is likely to be in error by considerable amounts just due to the inherent difficulties of conducting a national census.

35   It is easy to accurately measure the height of a structure with a reasonable degree of precision, say to the nearest 0.1 foot, but the given number has far too much precision, to the nearest 10 billionth of a foot, so the claim is not believable.

37   The number of college students is constantly changing with new enrollments, graduations, and dropouts, so the given number must be an estimate. The precision of the given number is probably unjustified. The number given may be approximately correct, but we would need more information to know if it is really accurate.

Section 2.3
Statistical Literacy and Critical Thinking

1   a)   The absolute change in the budget is \$5333 million - \$4919 million \$414 million.

   b)   The relative change in the budget is \$414 million/ \$4919 million = 0.084 or 8.4%.

   c)   The percentage change is (5185 - 5333)/5333 = -0.028 or -2.8%. Thus, there is a decrease of 2.8%.

d)   5% of $5333 million is 0.05 x $5333 million = $266.65 million. Thus, the new budget will be $5333 million - $266.65 million = $5066.35

3 There is a distinction to be made between percent and percentage points. The actual margin of error is 3 percentage points, meaning that we are 95% confident that the true percent is somewhere in the range from 22% to 28%. By saying that the margin of error is 3.0%, the statement (probably unintentionally) implies that the error can be up to 3.0% of 25%, or 0.75%. This would mean that the confidence interval had a much smaller range, from 24.25% to 25.75%.

5 The statement does not make sense. The **number** of people with cell phones may have increased by 1.2 million people, but 1.2 million people is not a percentage.

7 This makes sense. The 100% increase indicates that the loan rate doubled. For example, if the loan rate went from 4% to 8%, the increase was (8 - 4)/4 = 1.00 or 100%, and the rate was doubled.

Concepts and Applications

9 a)   75% is the same as 75/100 = 3/4 = 0.75 (or 75%)

b)   3/8 = 0.375 = 37.5%

c)   0.4 is the same as 4/10 or 2/5 = 0.40 = 40%

d)   80% is the same as 80/100 or 4/5 = 0.80 = 80%

11 a)   956 of 1348 pled guilty. This is 956/1348 = 0.71 or 71%

b)   392 + 58 = 450 were sent to prison. Thus 450/1348 = 0.33 = 33%, so 33% were sent to prison.

c)   392/956 = 0.41 = 41%, so 41% of those pleading guilty were sent to prison.

d)   58/72 = 0.81 = 81%, so 81% of those pleading not guilty were sent to prison.

13 (1387 - 2226)/2226 = -0.37 = -37%, so there is a 37% decrease in the number of daily newspapers from 1900 in the U.S.

15 (751183 - 634343)/634343 = 0.18 = 18%, so there is an 18% increase in passenger flights from 1996.

17 (2.09 million - 1.83 million)/ 1.83 million = 0.14 = 14%, so the *Wall Street Journal* circulation is 14% more than the *USA Today* circulation, or 114% of the *USA Today* circulation.

19 (67 million - 89 million)/89 million = -0.25 = -25%, so O'Hare handled 25% fewer passengers than Atlanta's Hartsfield Airport.

21 4.8% of 1385 = 0.048(1385) = 66, so 66 said that they do not make personal phone calls.

23 83% of 1005 = 0.83(1005) = 834, so 834 of those adults surveys reported having more than one television at home.

25 40% more than 100% is 140%, so the truck's weight is 140% of the car's weight.

27 The population of Montana is 20% less than 100% of the population of New Hampshire, so Montana's population is 80% of New Hampshire's population.

29 Yes. Reporting the error as 3% would imply that it was 3% of 89% or 0.03(0.89) = 0.0267 or 2.67 percent. The correct confidence interval is from 86% to 92%, whereas the implied confidence interval would be from 86.33% to 91.67%. In practical terms, there is not much difference in this case, but there is a very large difference when the reported percentage is small, say 11% instead of 89%.

31 The decrease in the percentage of high school seniors using alcohol is 68.2% - 52.7% = 15.5% or 15.5 percentage points. Since (52.7 - 68.2)/68.2 = -0.23 = -23%, there has been a 23% decrease from 1975 to now in the percent of high school seniors using alcohol.

33 The five-year survival rate for Caucasians for all forms of cancer increased

22 percentage points between the 1960s and the 1990s. This is a relative change of (new value – reference value)/ reference value = (61 – 39)/39 = 22/39 = 0.564 = 56.4% or an increase of 56.4%.

Section 2.4
Statistical Literacy and Critical Thinking

1   An index number is a ratio without any units, such as dollars. The number appears to be the actual cost of gasoline in 2011, not an index number.

3   Yes. The Consumer Price Index is based on the prices of goods, services, and housing, so if the prices in all three areas increase, the CPI must increase as well.

Concepts and Applications

5 $Index = \frac{Price\ Today}{Price\ 1980} = \frac{5.00}{1.22} = 4.098 = 409.8\%$

Thus the price index number for gasoline today is 409.8 with the 1980 price as the reference value.

7   The cost of gasoline in 1998 is the 1980 price multiplied by the 1998 price index expressed as a decimal or $1.22(0.9) = $1.10 per gallon.

9   To determine the gasoline price index using the 2000 price as the reference value, divide each of the other prices by the 2000 price and express the result as a percentage. For example, the 1960 price index is found from 0.31/1.56 = 0.199 = 19.9%, resulting in a price index of 19.9. The other price indices are found similarly.

Year	Price	Price as a Percentage of 2000 Price	Price Index (2000 = 100)
1960	$0.31	19.9%	19.9
1970	$0.36	23.1%	23.1
1980	$1.22	78.2%	78.2
1990	$1.23	78.8%	78.8
2000	$1.56	100.0%	100.0
2010	$2.84	182.1%	182.1

11   Using Table 2.1, which uses 1980 as the reference year, we see that the 2010 gasoline price index is 232.8. Since the gasoline costs are in the same ratio as the gasoline price indices,

$$\frac{Cost\ in\ 2010}{Cost\ in\ 1980} = \frac{Price\ Index\ in\ 2010}{Price\ Index\ in\ 1980}.$$

Therefore, $Cost\ in\ 2010 = (Cost\ in\ 1980)\frac{Price\ Index\ in\ 2010}{Price\ Index\ in\ 1980} = \$19.52\frac{232.8}{100.0} = \$45.44$

Note: Due to rounding in finding the gasoline price indices, this amount may be off by a penny or two.

13   The private college tuition in 2010 is $37000/5600 = 6.607 or 660.7% of the amount in 1980, but the CPI in 2010 is 218.1/82.4 = 2.647 or 264.7% of the CPI in 1980. Thus the cost of private college tuition rose much more than the cost of typical goods, services, and housing.

15 The median price in 2010 is 122000/75300 = 1.620 or 162.0% of the price in 1990 while the CPI in 2010 is 218.1/130.7 = 1.669 or 166.9% of the CPI in 1990. Thus the cost of homes in the South rose at a lower rate than the CPI.

17 $$Price\ in\ Miami = \frac{Index\ in\ Miami}{Index\ in\ Denver} X Price\ in\ Denver = \frac{194}{100}(300{,}000) = \$582{,}000$$

$$Price\ in\ Cheyenne = \frac{Index\ in\ Cheyenne}{Index\ in\ Denver} X Price\ in\ Denver = \frac{60}{100}(300{,}000) = \$180{,}000$$

19 $$Price\ in\ SF = \frac{Index\ in\ SF}{Index\ in\ Cheyenne} X Price\ in\ Cheyenne = \frac{382}{60}(250{,}000) = \$1{,}591{,}667$$

$$Price\ in\ Boston = \frac{Index\ in\ Boston}{Index\ in\ Cheyenne} X Price\ in\ Cheyenne = \frac{358}{60}(250{,}000) = \$1{,}491{,}667$$

Chapter 2 Review Exercises

1
a)   26% of 2303 is the same as 0.26(2303) = 599
b)   The data are discrete because only the counting numbers are used.
c)   1382/2303 = 0.60 or 60%
d)   53% of 1382 is the same as 0.53(1382) = 732
e)   Ratio. There is a true zero for ages.
f)   Nominal.

3 2,500,000,000,000/80,000,000,000 = 31.25, so the cost of health care increased by a factor of 31.25 or 3125%. The CPI increased by a factor of 218.1/44.4 = 4.91 or 491%, so the cost of health care increased at a rate much higher than that for goods, services, and housing.

Chapter 2 Quiz

1 The data are continuous since heights can take on any value within a given range.

3 The absolute error is 91.4 cm – 89.0 cm = 2.4 cm.

5 Nominal since states are just categories

7 5% of 1038 is 0.05 x 1038 = 52, so 52 of the respondents said that second-hand smoke is not at all harmful.

9 The index number for the second year is the ratio, in percent, of the second year net profit to that of the first year or 15257/12335 = 1.237 or 123.7%. Thus the index number for the second year is 123.7.

Section 3.1
Statistical Literacy and Critical Thinking

1 A frequency table is a table with two columns. One lists categories of data (the categories include all possible values of the data), and the other columns list the frequencies with which each category occurs in the data (number of data values in the category).

3 The cumulative frequency of a class is the sum of the frequencies for that class and those below it. Thus the cumulative frequencies are 4, 4 + 12 = 16, 4 + 12 + 16 = 32, 4 + 12 + 16 + 6 = 38, and 4 + 12 + 16 + 6 + 2 = 40.

5 This statement is not sensible since a frequency table must have a column of counts or frequencies (numbers of occurrences).

7 This statement is not sensible because each individual frequency must be a whole number. Cumulative frequencies are sums of those whole numbers and must therefore also be whole numbers.

Concepts and Applications

9

Grade	Frequency	Relative Frequency	Cumulative Frequency
A	3	0.100	3
B	10	0.333	13
C	11	0.367	24
D	2	0.067	26
F	4	0.133	30
Total	30	1.000	30

The Frequency column lists the number of times each grade occurred. The Relative Frequency column shows the portion of the time each grade occurred and the entries are obtained by dividing the Frequency by the total of 30. The Cumulative Frequency column gives the total number of grades that are equal to or better than the grade shown in the row.

11

Weight (Pounds)	Frequency	Relative Frequency	Cumulative Frequency
0.7900-0.7949	1	1/36	1
0.7950-0.7999	0	0	1
0.8000-0.8049	1	1/36	2
0.8050-0.8099	3	3/36	5
0.8100-0.8149	4	4/36	9
0.8150-0.8199	17	17/36	26
0.8200-0.8249	6	6/36	32
0.8250-0.8299	4	4/36	36
Total	36	1	36

The Frequency column lists the number of times each weight class occurred. The Relative Frequency column shows the portion of the time each class occurred and the entries are obtained by dividing the Frequency by the total of 40. The Cumulative Frequency column gives the total number of weights that are equal to or less than the upper class limit in the row.

13   As seen from the table below, the age category of 40 – 49 includes the most actors.

Age	Number of Actors
20-29	1
30-39	27
40-49	35
50-59	14
60-69	6
70-79	1

15   For Categories B and C, determine the frequencies by multiplying the total, 50, by 18% and 24% respectively, yielding frequencies of 9 and 12. Then add the frequencies for categories B, C, D, and F; subtract this total of 38 from the overall total of 50 to get the frequency for category A, which is 12. Now obtain each of the remaining relative frequencies by dividing the frequency by the total of 50 and converting to a percentage. For example, 12/50 = 0.24 = 24%.

Category	Frequency	Relative Frequency

A	12	24%
B	9	18%
C	12	24%
D	11	22%
F	6	12%
Total	**50**	**100%**

17 a) Adding up all of the frequencies, we find that the die was rolled 200 times.

b) The outcome was greater than 2 in 42 + 40 + 28 + 32 = 142 times.

c) Outcomes were 6 in 32 of the 200 rolls. 32/200 = 0.16 or 16%.

d) The relative frequencies are obtained by dividing each frequency by 200. Thus, the relative frequencies are 0.135, 0.155, 0.210, 0.200, 0.140, and 0.160.

19 a) and b)

Rating	Qwerty Frequency	Qwerty Relative Frequency	Dvorak Frequency	Dvorak Relative Frequency
0-2	20	38.5%	33	63.5%
3-5	14	26.9%	19	36.5%
6-8	15	28.8%	0	0.0%
9-11	2	3.8%	0	0.0%
12-14	1	1.9%	0	0.0%
Total	52	100%	52	100%

c) The Dvorak keyboard appears to be more efficient because it has more low ratings and fewer high ones.

Section 3.2
Statistical Literacy and Critical Thinking

1 The *distribution* of data is the way that the data values are spread over all data values.

3 By arranging bars from highest to lowest, the Pareto chart draws attention to the most important categories. The pie chart does not do that.

5 From the material presented, this statement does not make sense. Names of political parties are qualitative data, so a bar graph would be used to show the distribution of the parties. Histograms require quantitative data.

7 This statement is sensible. By positioning the categories with the highest frequencies at the left, the Pareto chart does draw attention to the most serious causes of defects.

Concepts and Applications

9 A histogram would work well to show the frequencies of the different categories of incomes, particularly if the number of scores is large. For a relatively small number of incomes, a stem-and-leaf diagram will also work well.

11   A time series diagram would be effective in showing any trend in the number of movie theaters.

13   a)

Category	Percentage
Popular fiction	50.4
Cooking/Crafts	10.2
General nonfiction	8.9
Religious	8.6
Psychology/Recovery	6.3
Technology/Science/Education	5.6
Art/Literature/Poetry	3.7
Reference	2.6
All other categories	2.5
Travel/Regional	1.3

b)   Pareto Chart

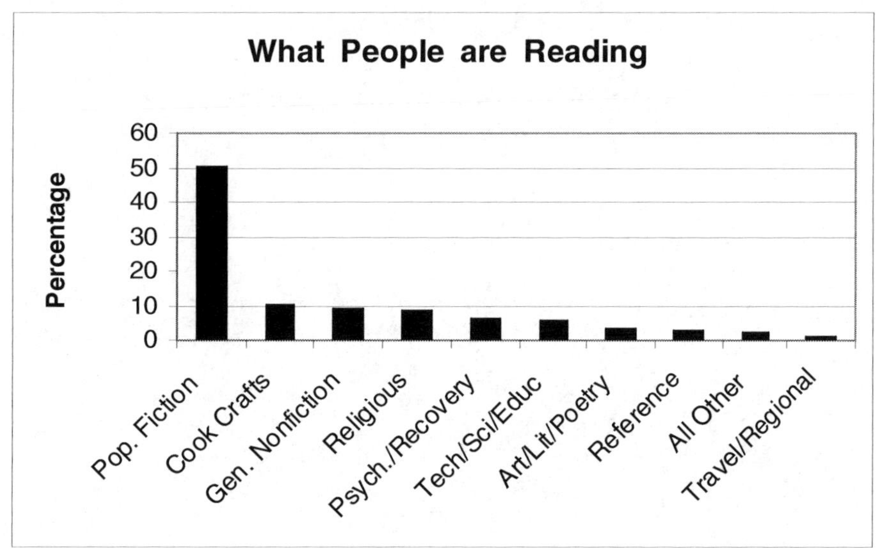

c)   The Pareto chart is easier to read than the pie chart when there are many categories with small percentages. It also makes it easier to see which categories are the most popular.

15

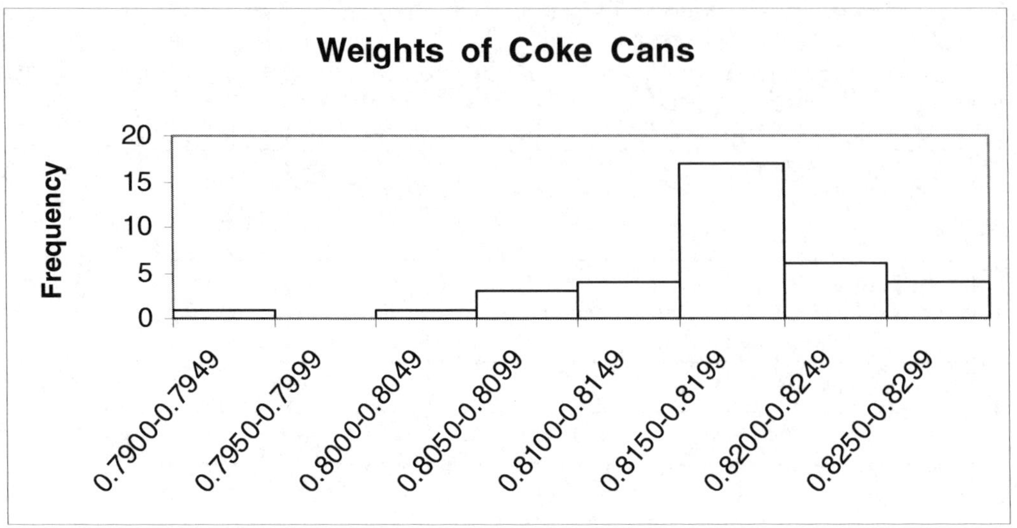

17

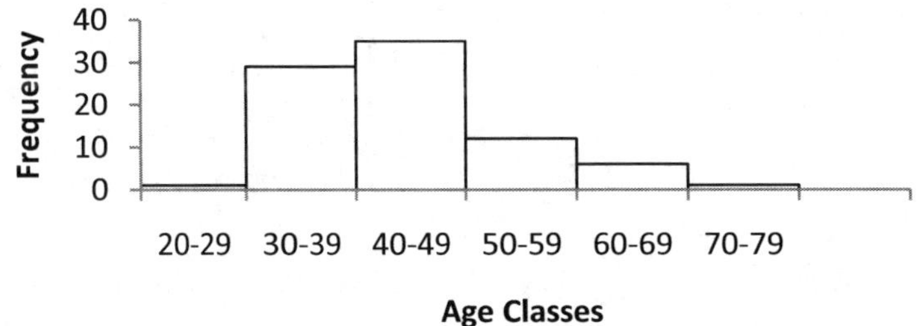

19

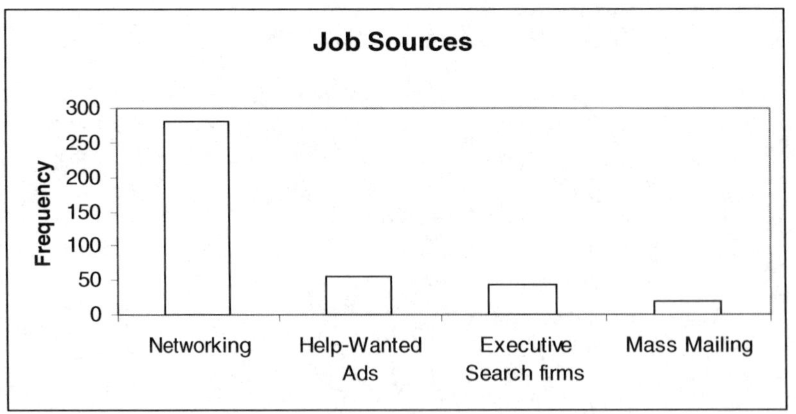

Clearly, networking is the best strategy.

21

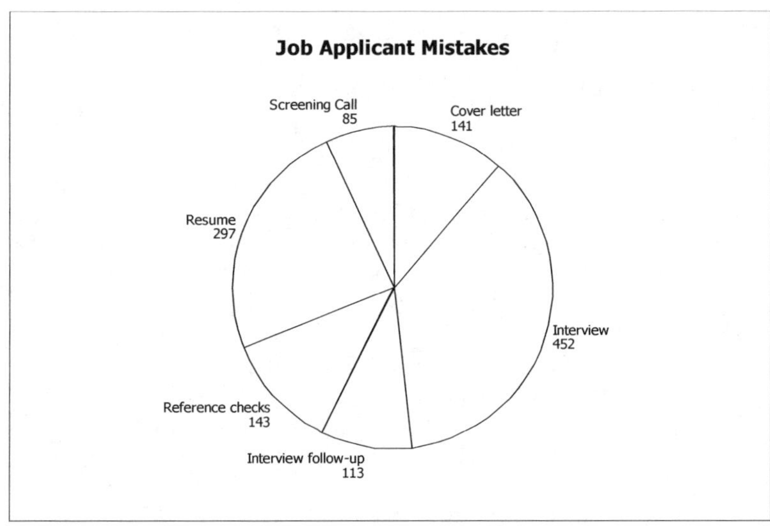

23

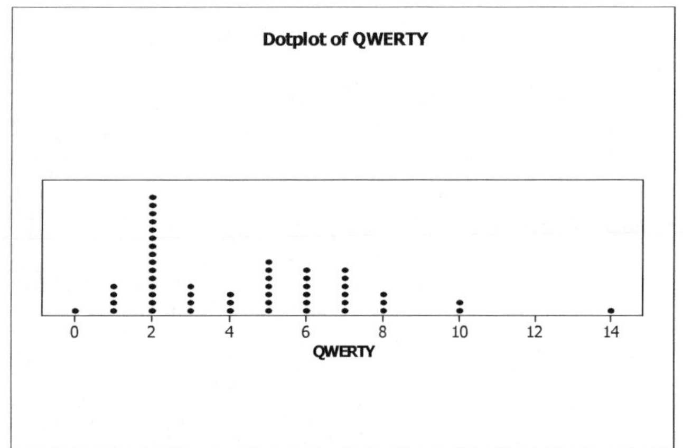

25

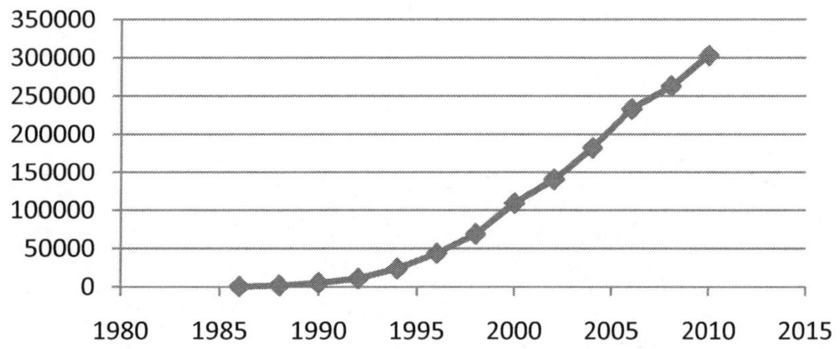

The graph does not show linear growth.

27

Stem	Leaves
6	7
7	25
8	5889
9	09
10	0

The lengths of the rows are similar to the heights of bars in a histogram, so that longer rows of data correspond to higher frequencies.

Section 3.3
Statistical Literacy and Critical Thinking

1   Yes, any histogram can be depicted as a three-dimensional histogram (as in Figure 3.18). The three-dimensional version of a histogram does provide more visual appeal, but it does not provide any additional information.

3   Geographical data are raw data corresponding to different geographic locations. Two examples of displays of geographical data are color-coded maps and contour maps.

5   The statement does not make sense. The production costs are one-dimensional, so a time series plot is appropriate.

7   The statement does make sense. However, a plain bar graph might be better because the bases of the bars would all be at the same level, making comparisons easier. The proposed graph would be virtually unreadable if many cities were included.

Concepts and Applications

9   a)   Numbers of females consistently outnumber the numbers of males. The numbers of both genders are increasing gradually over time.

  b)

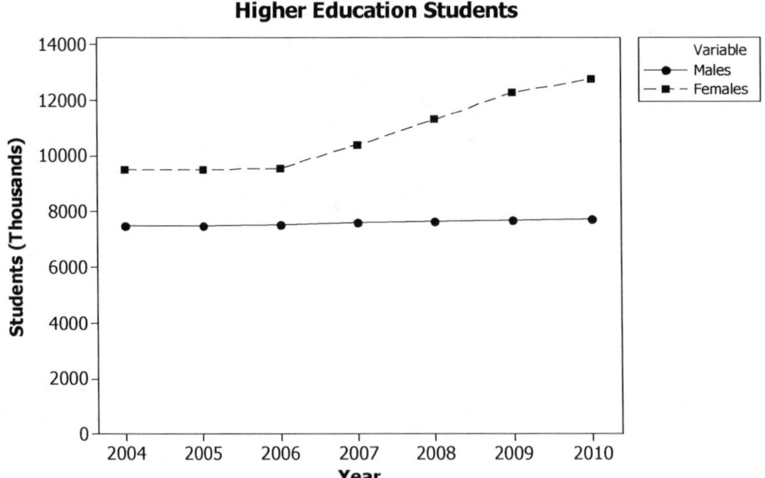

The line graph makes the comparisons much easier since one does not have to subtract the numbers of females from the totals to get the numbers of males. The line graph shows that the rate of growth for females is slightly greater than that of males, but that feature is very difficult to see in the stacked plot. The line graph is also less cluttered and is therefore easier to understand and interpret.

11 a) Males consistently have much higher median incomes than females, and both males and females have steadily increasing incomes over time.

b)

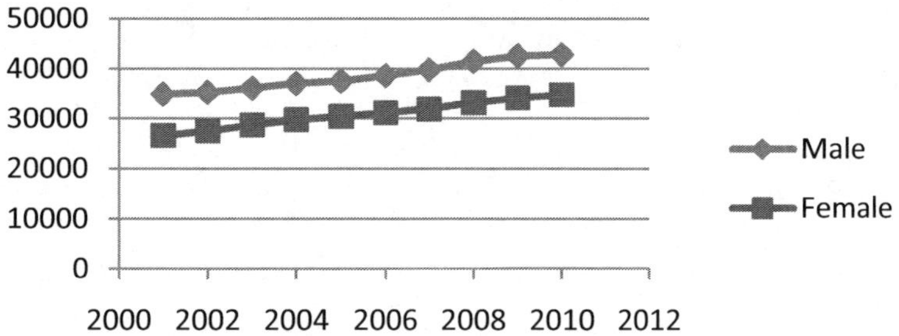

The line graph makes it easier to examine the trend over time. Also the line graph is less cluttered, so that the information is easier to understand and interpret.

13 This stack plot shows that entitlement spending (Social Security, Medicare, Medicaid, and insurance subsidies) is projected to grow dramatically in coming decades. The graph shows that, without an increase in tax revenues, this entitlement spending would consume all government revenue by mid-century, which would leave no funds for other government programs, including the military. The clear message is that without changes, entitlement spending is on a track that would cause severe budgetary problems.

15 The most obvious trend is that the rates of melanoma deaths are generally higher in the southern and southwestern states than in the states north of them. This might be the result of people in these states spending more time outdoors, exposed to sunlight. As a researcher, you might be particularly interested in regions that deviate from general trends. There are some very low rate counties along the Mexican border in Texas, New Mexico, Arizona, and California; and there are a few high rate counties in northern Minnesota, North Dakota, and Montana. These counties are bucking the overall trend and would therefore warrant some special research attention to find out why they are different from the counties surrounding them.

17 Since the primary purpose of graphing this data will be to show a comparison of the number of fatalities involving no alcohol, moderate alcohol, and high alcohol, a line graph with one line for each category of fatalities will make the comparisons easiest. While the number of fatalities involving no alcohol has increased from 1990 to 2005 and then fell dramatically in 2010, the number of fatalities involving moderate alcohol has decreased slightly and the number involving high alcohol has declined significantly.

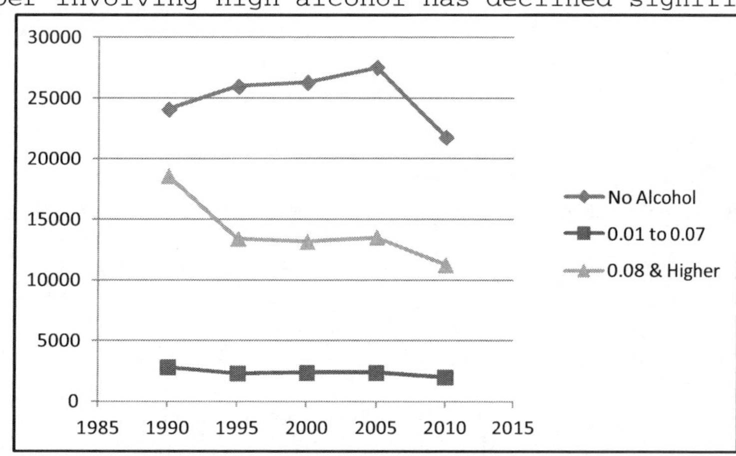

19 Data of this type can be displayed (in theory) by stacked bar graphs with a separate bar for each country. Such a graph would display the *numbers* of deaths in each country that occurred by firearm homicides, suicides, and accidents, but any interpretations must take into account the vast differences in the populations of the countries listed. Unfortunately, the numbers for the U.S. are so much larger than for the countries at the bottom of the list that no graph can convey the information in the table as well as the table can. To aid in interpretation of this information, the numbers should be presented in fatalities per 100,000 of population.

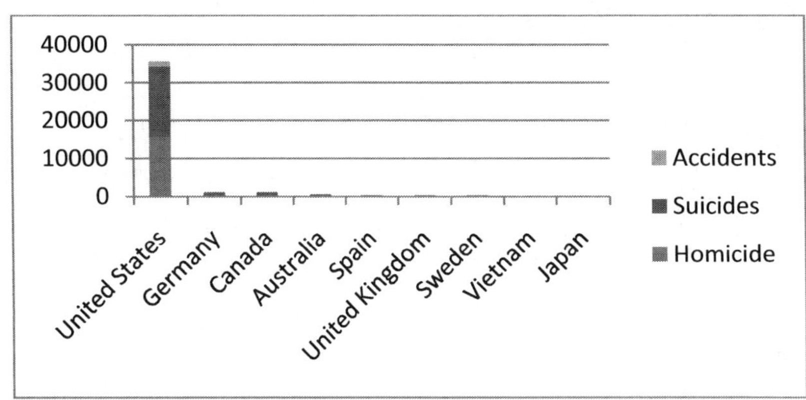

Section 3.4
Statistical Literacy and Critical Thinking

1 A difference of 1 foot is not meaningful. By drawing a bar graph with the vertical scale starting at a value such as 130 feet, the difference will be greatly exaggerated.

3 The scale is called an exponential scale. The advantage of such a scale is that it enables us to include values that vary over a very large range.

Concepts and Applications

5 The fuel consumption for the Aveo is about 34 mpg and for the Civic, it is about 37 mpg. The graphic makes the difference look deceptively large because it starts at the bottom from 30 mpg rather than from zero. A fairer representation would begin from zero.

7 By using objects that have volume, the graph is misleading by creating the perception that in comparison to the oil consumption in Japan, the U.S. consumption is much larger than it actually is. Comparing the actual amounts, we see that the U.S. consumption of 20.0 million barrels per day is roughly four times Japan's consumption of 5.4 million barrels per day. By using a barrel that is about four times as wide, four times as tall, and four times as deep, the larger barrel has a volume that is roughly 64 times that of the smaller barrel, instead of the correct amount of four times.

9 a) Because of the 3-d appearance of the pie charts, the sizes of the wedges on the page do not match the percentages. Instead, they show how the wedges would look if the entire pie were tilted at an angle. This distortion makes it difficult to see the true relationships among the categories.

  b) Because of the 3-d appearance of the pie charts, the sizes of the wedges on the page do not match the percentages. Instead, they show how the wedges would look if the entire pie were tilted at an angle. This distortion makes it difficult to see the true relationships among the categories.

c)   The pie charts below are not distorted by a 3-d view.

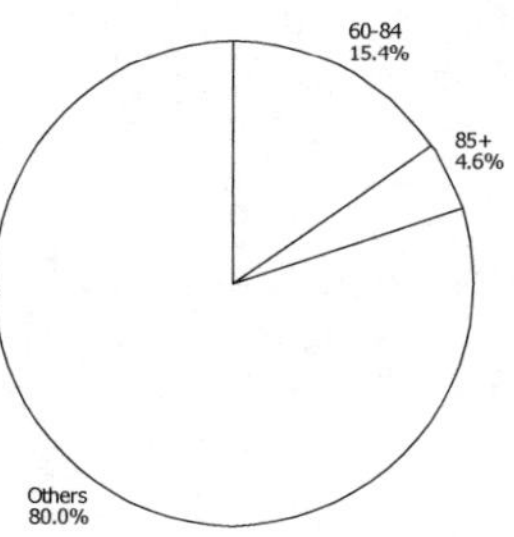

d)   Both the 65-84 and 85+ populations are expected to be a larger percentage of the total population in 2050 than they were in 1990.

11   The percent change in the CPI is greatest in 1990. In 2009, the percent change in the CPI is negative, so prices decreased from those in 2008. Prices have increased in every year except 2009, but the increases near the end of the time period are lower than those near the beginning.

13 The actual minimum wage in unadjusted dollars has either remained constant or risen steadily since 1955. The purchasing power increased from 1955 to 1968, then decreased from 1968 to 1989, and it has been relatively stable in recent years. The purchasing power in 2011 is just slightly higher than it was in 1955.

Chapter 3 Review Exercises

1   a)

Regular	Pepsi
Bin	Frequency
0.8130-0.8179	5
0.8180-0.8229	12
0.8230-0.8279	12
0.8280-0.8329	5
0.8330-0.8379	0
0.8380-0.8429	2
Total	36

b)

Diet	Pepsi
Bin	Frequency
0.7740-0.7779	5
0.7780-0.7819	5
0.7820-0.7859	16
0.7860-0.7899	8
0.7900-0.7939	2
Total	36

c)   The weights of the regular Pepsi cans are consistently larger than those of diet Pepsi. The volumes of the two types of cans are likely to be the same, but the regular Pepsi cans are heavier due to the sugar in those cans that is not in the cans of diet Pepsi.

3   a)

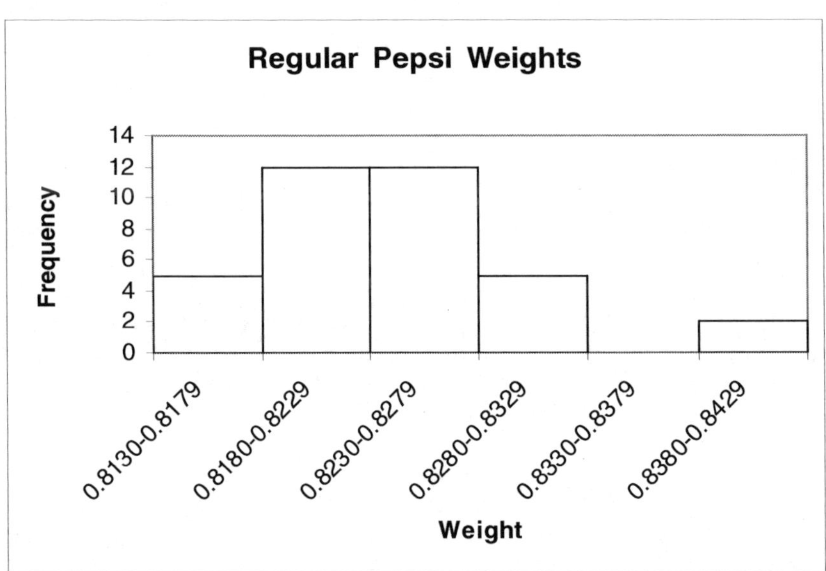

b)

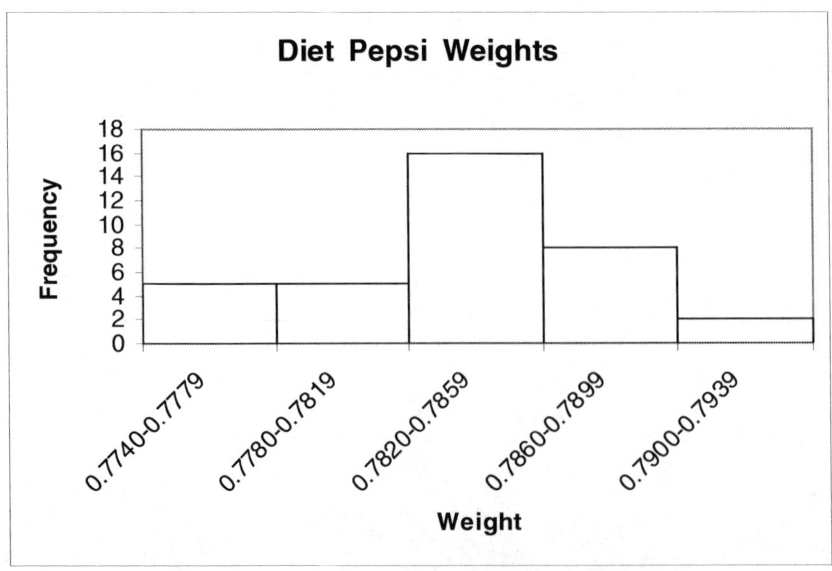

c)   The shapes of the histograms are similar, higher in the center and
lower at the sides. However, the range of values is very different,
indicating that the weights of the cans of regular Pepsi are
considerably higher than those of diet Pepsi.

5   For the Pareto Chart, use the data from Exercise 4, but rearrange the rows
of the table so that the frequencies are in decreasing order. Then
highlight the entire table and select **Insert – Column** (from the charts) and
click on the first 2-D graph type. Again the graph can be modified by right
clicking on various areas of the graph and changing the appearance of the
graph.

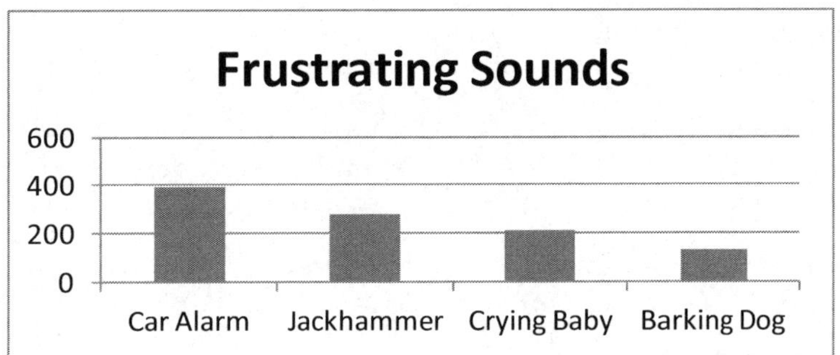

By drawing attention to the most frustrating sounds, the Pareto Chart is
more effective. By using a meaningful scale, the Pareto Chart also does a
better job of showing the relative importance of these frustrating sounds.

Chapter 3 Quiz

1   A histogram is the most appropriate since there is only one variable (IQ)
and it is quantitative. A bar graph is used for qualitative data, and both
the multiple bar graph and stack plot are used only when there is more than
one variable.

3   The values (in minutes) are 61, 62, 62, 62, 62, 67, 69.

5   The "0.25" indicates that, among all of the data values, 0.25 or 25% of them
fall between 20 and 29, inclusive.

7   By using a vertical scale that does not start at zero, the difference
between the two frequencies is exaggerated.

9   The values are 0, 1, 1, 10, 11, 12, 49, 49.

CHAPTER 4 ANSWERS

<u>Section 4.1</u>
Statistical Literacy and Critical Thinking

1   No. The mean specifically refers to the sum of the data values divided by the number of data values. The term "average" could refer to any of several statistics, including the mean, median, and mode.

3   Yes. The executive secretary's income is much higher than that of the other 23 students, so it is an outlier. In general, the definition of an outlier is somewhat vague, so an outlier cannot be identified clearly and objectively.

5   This statement does not make sense. The numbers on the jerseys are just substitutes for the players' names. Finding an average of names has no meaning.

7   This statement is sensible. It is possible for a set of data to have the same values for the mean, median, and mode. For example, the data set consisting of 4, 6, 6, 6, 8 has mean, median, and mode all equal to 6.

9   The mean best describes average heights of professional basketball players since there are only a few heights that would be considered extreme and even those would not be very extreme. The mean is also better because it takes every height into account.

11   The median would best describe these data since he exact values of the ages in the "over" categories are not known, and thus the mean cannot be computed.

Concepts and Applications

13   For the mean, total the 10 numbers and divide by 10. Round the answer to one more decimal place than shown in the data.
$$Mean = \frac{Total\ of\ Values}{Total\ number\ of\ Values} = \frac{51+63+36+43+34+62+73+39+53+79}{10} = 53.3$$
For the median, first put the ten numbers in increasing order. Since there is an even number of data values, the median is the average of the middle two (fifth and sixth) numbers. Thus median = (51 + 53)/2 = 52.0.
The mode is the number that occurs most often. Since no number occurs more than once, there is no mode.

15   For the mean, total the eight numbers and divide by 8. Round the answer to one more decimal place than shown in the data.
$$Mean = \frac{Sum\ of\ the\ Numbers}{Total\ number\ of\ values} = \frac{53+52+75+62+68+58+49+49}{8} = 58.3$$
For the median, first put the eight numbers in increasing order. Since there is an even number of data values, the median is the average of the middle two (fourth and fifth) numbers. Thus median = (53 + 58)/2 = 55.5.
The mode is the number that occurs most often. Since 49 occurs twice and no other number occurs more than once, 49 is the mode.

17   For the mean, total the 12 numbers and divide by 12. Round the answer to one more decimal place than shown in the data.
$$Mean = \frac{Total\ of\ Values}{Total\ number\ of\ Values} = \frac{0.27+0.17+0.17+0.16+0.13+0.24+0.29+0.24+0.14+0.16+0.12+0.16}{12} = 0.188$$
For the median, first put the twelve numbers in increasing order. Since there is an even number of data values, the median is the average of the two middle (sixth and seventh) numbers. Thus median = (0.16 + 0.17)/2 = 0.165.
The mode is the number that occurs most often. Since 0.16 occurs three

19 times and no other number occurs more than twice, 0.16 is the mode. For the mean, total the 11 weights and divide by 11. Round the answer to one more decimal place than shown in the data. $Mean = \dfrac{Total\ of\ Values}{Total\ number\ of\ Values} =$

$$\frac{0.957+0912+0.842+0.925+0.939+0.886+0.914+0.913+0.958+0.947+0.920}{11} = 0.9194$$

For the median, first put the eleven numbers in increasing order. Since there is an odd number of data values, the median is the middle (sixth) number. Thus median = 0.9200 g.
The mode is the number that occurs most often. Since no value occurs more than once, there is no mode.

21 a)   For the mean, total the seven areas and divide by 7.
     The sum of the seven areas is 1,103,100 square miles, so the mean is 1103100/7 = 157,586 square miles.
     For the median, first put the seven numbers in increasing order. Since there is an odd number of data values, the median is the middle (fourth) number. Thus median = 104,100 square miles.

   b)   Alaska is an outlier on the high end. Without Alaska, the mean is 487,900/6 = 81,317 square miles. The median is the average of the third and fourth values = (53200 + 104100)/2 = 78650 square miles.

   c)   Connecticut is an outlier on the low end. Without Connecticut (but with Alaska), the mean is 1,097,600/6 = 182,933 square miles. The median is the average of the third and fourth values = (104100 + 114000)/2 = 109050 square miles.

23 a) $Mean = \dfrac{Total\ of\ Values}{Total\ number\ of\ Values} = \dfrac{80+84+87+89}{4} = 85.0$

   b)   Since the mean will be the total divided by five, the total will need to be 5 x 88 = 440 in order for the mean to be 88 after the next quiz. Since the total is already 340 after the first four quizzes, the next quiz score will need to be 440 – 340 = 100.

   c)   If you achieve a score of 100, your mean score will be (340+100)/5 = 440/5 = 88. Thus it's not possible to have a mean score higher than 88 after the next quiz.

25 The mean score of your students is 523/7 = 74.7. The median score is 70. If the district "average" is a mean, then your students are above average; if the district average is a median, then your students are below average.

27 The mean weight of all of the peaches is the total weight divided by the total number of peaches or (18 + 22 + 24) pounds/(50 + 55 + 60) = 64/165 = 0.39 pounds.

29 Each student is taking three classes with enrollments of 20 each and one class with an enrollment of 100, so the mean size of each student's classes is 160/4 = 40. There are three classes with 100 students each and 45 classes with 20 students each, making a total enrollment of 1200 students in 48 classes. Thus the mean enrollment per class is 1200/48 = 25.

31 Batting average = total hits/total at-bats = 5/12 = .417. This is the average number of hits per at-bat.

33 Probably not. The overall rate of defects would be 3% only if both sites produced exactly the same number batteries.

35 If a yes vote is valued at 1 and no vote is valued at 0, then the outcome is (400x1 + 600x0)/1000 = 0.4. Thus the average number of yes votes per share is less than 0.5 and the result is a No vote.

37 Since the data are at the nominal level, it makes no sense to find a mean or a median. The mode is 1 and it correctly indicates that the smooth-yellow phenotype is the most common of the four phenotypes.

Section 4.2
Statistical Literacy and Critical Thinking

1   The distribution is not symmetric.

3   Because the professors have successfully completed a rigorous program of education, their IQ scores are likely to all be above the average for the general population and closer together, so they probably have less variation than the IQ scores of randomly selected adults. The lower variation of IQ scores of the professors results in a graph with less spread than the graph of IQ scores from the randomly selected adults.

5   This does not make sense. A data distribution may have three modes and still be symmetric.

7   This statement makes sense since it is possible for a unimodal distribution to be skewed.

Concepts and Applications

9   The distribution has two modes, is skewed left, and has wide variation.

11   The distribution has one mode, is nearly symmetric, and has moderate variation.

13   a)   The distribution is not symmetric and is right-skewed.
 b)   About half of the salaries (409) are greater than or equal to the median.

15   a)   One mode at $0
 b)   Right-skewed, because there will be many students making little or nothing

17   a)   One mode
 b)   Symmetric

19   a)   One mode
 b)   Symmetric

21   a)   One mode
 b)   Left-skewed

23   a)   One mode
 b)   Right-skewed

25   a)   One mode
 b)   Right-skewed

Section 4.3
Statistical Literacy and Critical Thinking

1   The range is found by subtracting the lowest value from the highest value. It is a measure of variation. A major disadvantage of the range is that its value depends only on the highest and lowest values and does not take into account the rest of the data.

3   The statement is incorrect because it implies that the standard deviation depends only on the high and low values, but the standard deviation uses every data value.

5   This does not make sense. The first quartile is the same as the 25th percentile. Only if there were a very large number of scores equal to Jennifer's could the 25th and 35th percentiles be equal.

7   This makes sense. The median and the second quartile are the same.

Concepts and Applications

9   The range is 75 − 34 = 45.
 Enter the data values in Excel in A1 through A10. In A11, enter the formula =STDEV(A1:A10). The result is 15.7 (rounded to one decimal place).

10   The range is 381 − 0 = 381.0
Enter the data values in Excel in A1 through A15. In A16, enter the
formula =STDEV(A1:A15). The result is 105.9 (rounded to one decimal
place).

11   The range is 75 − 49 = 26.0
Enter the data values in Excel in A1 through A8. In A9, enter the
formula =STDEV(A1:A8). The result is 9.5 (rounded to one decimal
place).

13   The range is 0.29 − 0.12 = 0.17
Enter the data values in Excel in A1 through A12. In A13, enter the
formula =STDEV(A1:A12). The result is 0.057 (rounded to three decimal
places).

15   The range is 0.958 − 0.842 = 0.1160
Enter the data values in Excel in A1 through A11. In A12, enter the
formula =STDEV(A1:A11). The result is 0.0336 (rounded to four decimal
places).

17   For *Cat on a Hot Tin Roof,* Range = 11 − 1 = 10.0 and the standard
deviation equals 2.6, found by using Excel. For *Cat in the Hat*, Range
= 5 − 2 = 3.0 and the standard deviation equals 0.9. By either
measure, there is much less variation in *Cat in the Hat*.

19   The one day range = 8 − (−3) = 11 degrees and the standard deviation is
2.6 degrees. The five day range = 6 − (−9) = 15 degrees and the
standard deviation is 4.5 degrees. As expected, the one day forecasts
show less variance in the errors than do the five day forecasts.

21   a)   Percentile = (25/465) x 100 = 5
       b)   Percentile = (322/465) x 100 = 69
       c)   Percentile = (224/465) x 100 = 48

23   a)

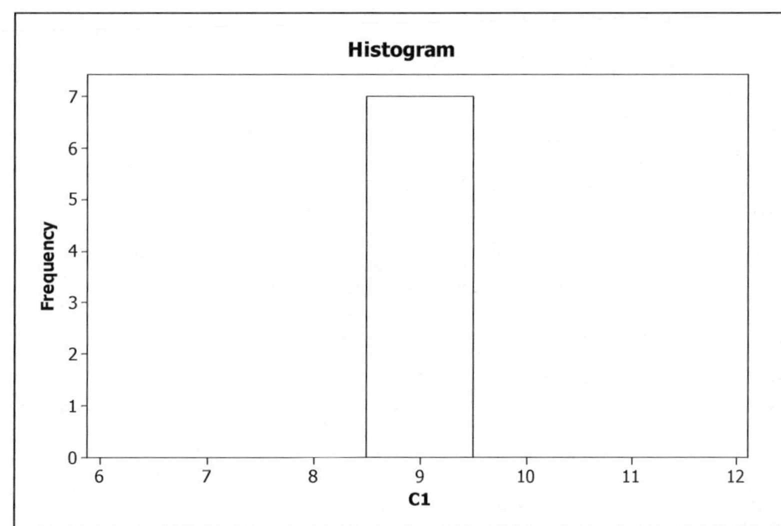

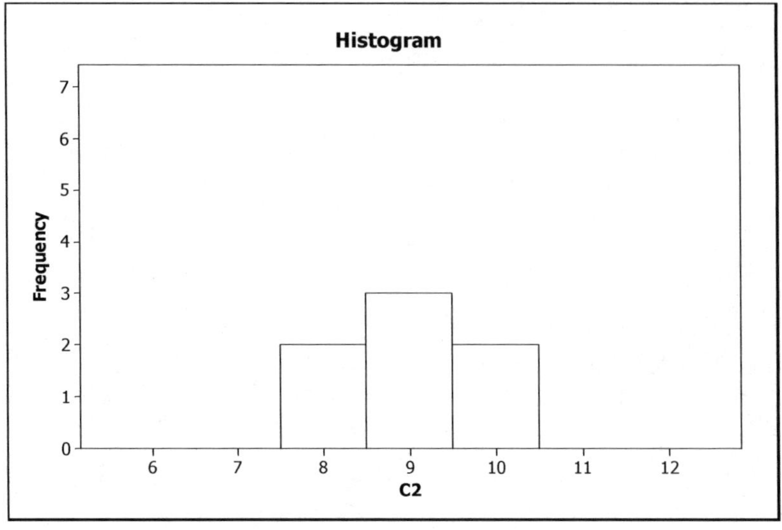

Histogram
Frequency
C2

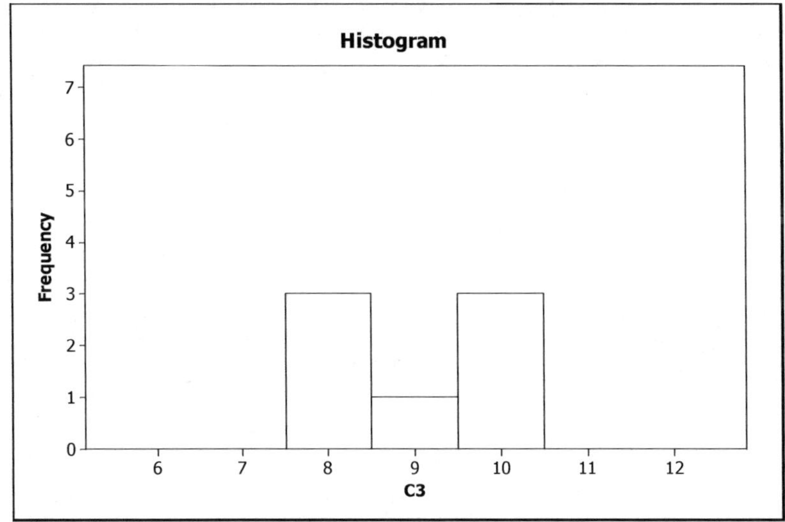

Histogram
Frequency
C3

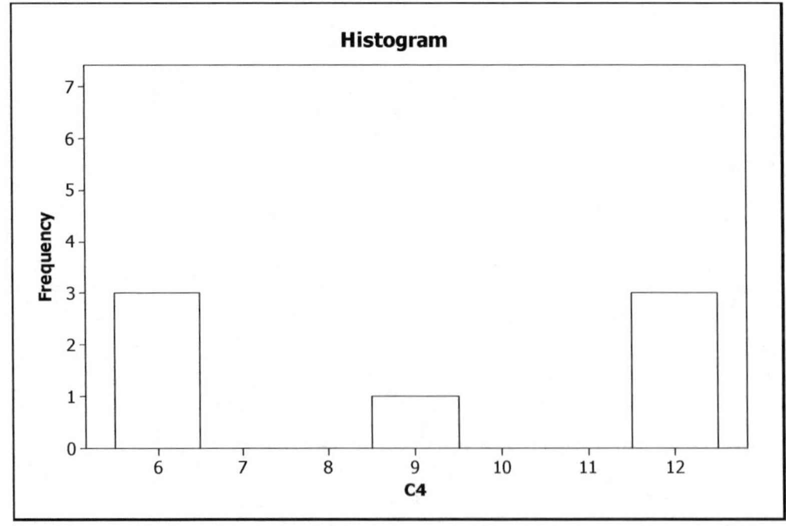

Histogram
Frequency
C4

b)

	Minimum	Q1	Median	Q3	Maximum
Set 1	9	9	9	9	9
Set 2	8	8	9	10	10
Set 3	8	8	9	10	10
Set 4	6	6	9	12	12

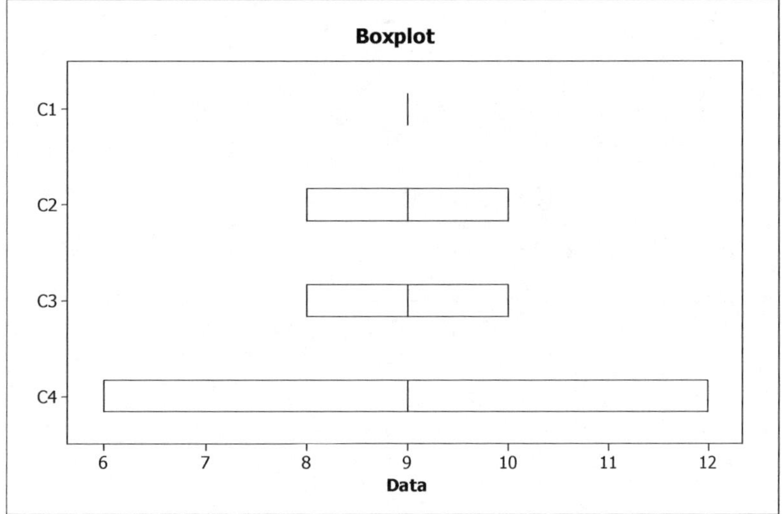

c)   Using Excel, enter each of the data sets in a different column and use the STDEV function. The standard deviations are 0.0, 0.8, 1.0, and 3.0 respectively for the four sets of data.

d)   The standard deviation takes every value into account and thus distinguishes the variation in the four sets. Note that the range would be the same for the second and third sets even though the sets are different.

25   a)

	Mean	Median	Range
Faculty	2.0	2.0	4.0
Students	6.18	6.0	9.0

b)

	Low	Q1	Median	Q3	High
Faculty	0	1	2	3	4
Students	1	4	6	9	10

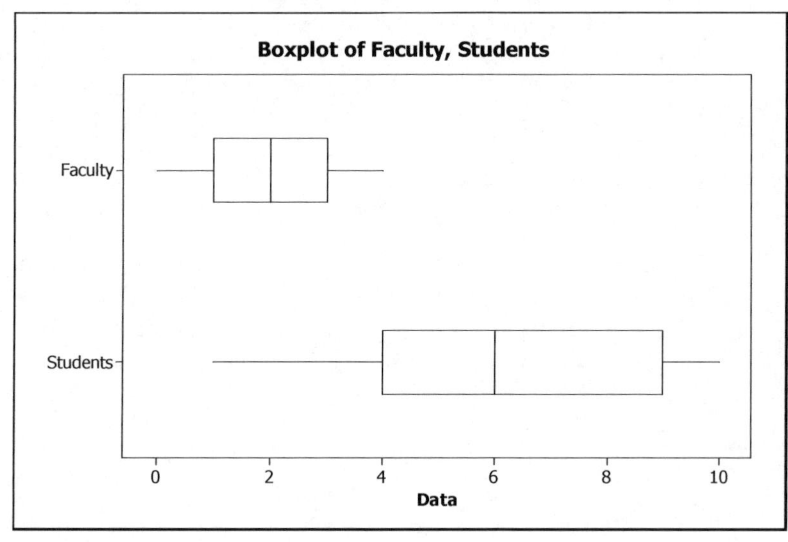

c)   Using Excel, we find that the standard deviations are 1.2 years for the faculty and 3.0 years for the students.

d)   Using the Range Rule of Thumb, we estimate the standard deviation for the faculty to be Range/4 = 4/4 = 1.0 and for the students to be Range/4 = 9/4 = 2.3. These Range Rule of Thumb estimates are reasonably close to the actual standard deviations considering that the number of data values is very small.

e)   The faculty tend to have newer cars and there is less variation in the ages of their cars than for the students.

27 a)

	Mean	Median	Range
First 7	58.3	57.0	4.0
Last 7	56.1	54.0	23.0

b)

	Low	Q1	Median	Q3	High
First 7	57	57	57	61	61
Last 7	46	47	54	64	69

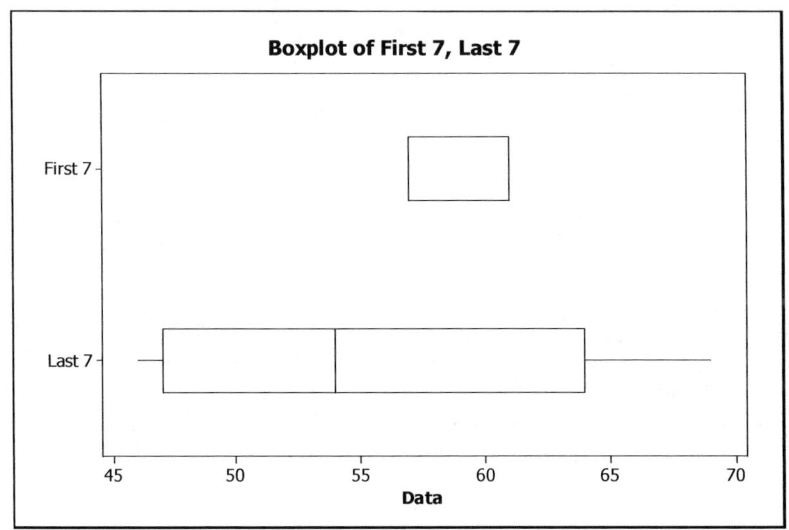

c)   Using Excel, we find that the standard deviations are 1.9 years for the first 7 and 8.7 years for the last 7.

d)   Using the Range Rule of Thumb, we estimate the standard deviation for the first 7 to be Range/4 = 4/4 = 1 and for the last 7 to be Range/4 = 23/4 = 5.75. These Range Rule of Thumb estimates are reasonably close (but both low) to the actual standard deviations considering that the number of data values is very small.

e)   The means and medians are quite close for the two groups of presidents, but there is much more variation in ages for the last 7 than for the first 7.

29   The second site has better quality. Although the second site has a mean that is a little farther from the target of 12 volts, the standard deviation is much less, so the voltages will be more consistent.

31   Having more stocks in the portfolio reduces the variation in the returns, because some stocks are likely to go up and others go down. The reduced variation means reduced risk, though also a lower likelihood of large gains.

Section 4.4
Statistical Literacy and Critical Thinking

1   A false positive occurs when the test indicates use of banned substances for someone who does not actually use them. A false negative occurs when the test indicates that banned substances are not used by someone who actually does use them. A true positive occurs when the player tests positive and he actually does use banned substances. A true negative occurs when the player tests negative and he does not use banned substances.

3   True negative.

5   This statement makes sense. No matter how the assignments and exams are weighted, a person who has a higher mean score on both will have a higher overall mean score.

7   This statement does not make sense. A positive test result indicates the presence of the disease, so the patient should not be happy.

Concepts and Applications

9   Josh had the higher batting average in each half of the season. Overall, Josh had 85 hits in 220 at-bats (.386 average), while Jude had 80 hits in 200 at-bats (.400 average). Thus Jude had the higher overall average. Josh had the higher average in each half, but Jude had the higher overall average.

11 a)   New Jersey has the higher average in both racial categories, but Nebraska has the higher overall average.

b)   This can happen if Nebraska has a high percentage of whites and New Jersey has a high percentage of non-whites.

c)   Nebraska's overall average is .87x281 + .13x250 = 277.

d)   New Jersey's overall average is .66x283 + .34x282 = 272.

e)   Simpson's Paradox occurs because New Jersey has the higher average in each racial category, but Nebraska has the higher overall average.

12 a)   The SAT scores in all five grade categories declined between 1988 and 1998.

b)   The overall SAT scores increased between 1988 and 1998.

c)   Even though the average SAT scores declined in each grade category, the overall average increased. This is an illustration of Simpson's Paradox, resulting from the changing percentages of students who were in each grade category, with higher percentages of students in the three "A" categories and lower percentages in the "B" and "C" categories. The total number of students is also probably different in the two years and this also contributes to the paradox.

13 a,b) Each death rate is computed by divided the number of deaths by the population. For New York whites, this is 8400/4675000 = 0.0018. The rest are shown in the table below.

Race	New York	Richmond
White	0.0018	0.0016
Non-white	0.0054	0.0034
Total	0.0019	0.0023

c)   The rate for both whites and nonwhites was higher in New York than in Richmond, yet the overall rate was higher in Richmond than in New York. The percentage of nonwhites was significantly lower in New York than in Richmond.

15   a)   Spelman has a better record for home games (34.5% vs. 32.1%) and away games (75.0% vs. 73.7%), individually.

     b)   Morehouse has a better overall average, 65 wins and 39 losses, against 22 wins and 23 losses for Spelman.

     c)   Morehouse has the better team since teams are generally rated on overall records.

17   a)   Fifty-seven appeared to be lying. Forty-two were actually lying and 15 of them were telling the truth. Fifteen out of 57 or 26% of those who appear to be lying were not actually lying.

     b)   Forty-one appeared to be telling the truth. Thirty-two of them were actually telling the truth and 9 were lying. Thirty-two out of 41 or 78% of those who appeared to be telling the truth were actually truthful.

19   A higher percentage of female applicants (20%) were hired for white-collar positions than of male applicants (15%). A higher percentage of female applicants (85%) were hired for blue-collar positions than of male applicants (75%). When white-collar and blue-collar positions are combined, a higher percentage of male applicants were hired (330/600 = 55%) than of female applicants (125/300 = 41.7%). Since qualifications for the two types of positions are different, it makes no sense to combine the two categories.

21   a)   In the general population 57 + 3 = 60 out of 20,000 are infected. This is an incidence rate of 6/20000 = 0.003 or 0.3%. In the at-risk population, 475 + 25 = 500 out of 5000 are infected. This is an incidence rate of 500/5000 = 0.1 = 10%. In the general population, the detection rate is 475/500 = 0.95 = 95%. For the at-risk population, the detection rate is 57/60 = 0.95 = 95%.

     b)   In the at-risk population, 475 out of 500 infected with HIV test positive (95%). There are 475 + 225 = 700 who test positive. Of these 475 have HIV (475/700 = 0.679 or 67.9%). The two numbers are different because they measure different things. The 700 who test positive include 225 who were false positives. While the test correctly identifies 95% who have HIV, it also incorrectly identifies 225 who do not. Thus only 67.9% of those who test positive actually have the disease.

     c)   In the at-risk population, a person who has tested positive has about a 67% chance of having HIV. This is nearly seven times a great as the 10% incidence rate for the general population. The test is very valuable in identifying those with HIV.

     d)   In the general population group, patients with HIV test positive 57times out of 60, a 95% rate. Of those who test positive, 57 out of 57 + 997 = 1054 were infected with HIV. This is a rate of 57/1054 = 0.054 or 5.4%. The two numbers are different because they measure different things. The 1054 who test positive include 997 who were false positives. While the test correctly identifies 95% of those with HIV, it also incorrectly identifies some who do not have HIV. Thus only 5.4% of those who test positive actually have HIV.

     e)   In the general population, a patient who tests positive for the disease has about a 5.4% chance of actually having the disease. This is 18 times as great as the incidence rate (0.3%) of the disease in the general population. Thus the test is very valuable in identifying those with HIV.

Chapter 4 Review Exercises

1   a)   Non-filtered: Using Excel, Mean is 1.28 mg and median is 1.15 mg. Filtered: Mean is 0.93 mg and median is 1.00 mg.

    b)   Non-filtered: Using Excel, Range is 0.70 mg and standard deviation is 0.26 mg. Filtered: Range is 0.80 mg and standard deviation is 0.24 mg.

    c)   The five-number summary for Non-filtered is 1, 1.1, 1.15, 1.4, 1.7. For Filtered, it is 0.4, 0.8, 1.0, 1.1, 1.2.

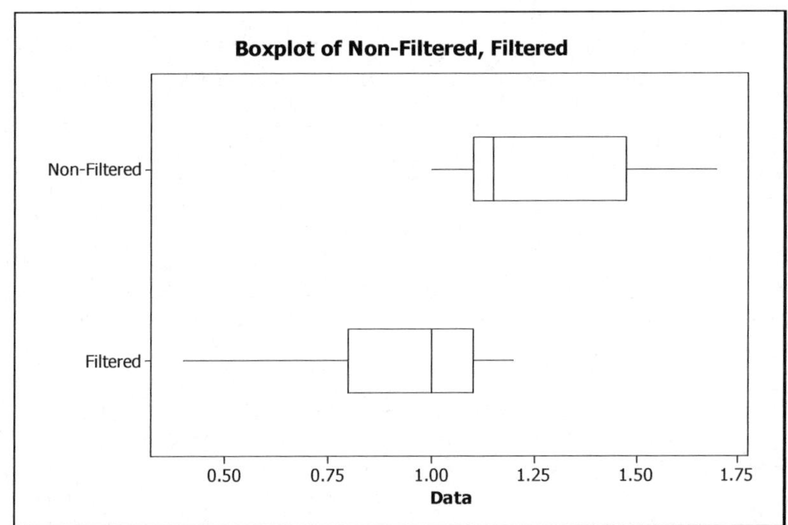

    d)   Non-filtered: The standard deviation is estimated to be 0.7/4 = 0.18 mg and it is actually 0.26 mg. Filtered: Standard deviation is estimated to be 0.80/4 = 0.20 and it is actually 0.24 mg. Both estimates are in the general ballpark of the actual values.

    e)   There does appear to be a difference. The nicotine in filtered cigarettes appears to be generally lower, so the filters appear to be at least somewhat effective.

3   a)   If all the values are the same, there is no variation and the standard deviation is aero.

    b)   Although both batteries have the same mean lifetime, the batteries with the smaller standard deviation are better because their lifetimes will be closer to the mean, and fewer of them will strand drivers by failing sooner than expected.

    c)   The outlier will pull the mean up or down, depending on whether it is larger or smaller than the other values.

    d)   The outlier will not affect the median.

    e)   The outlier will increase the range.

    f)   The outlier will increase the standard deviation

Chapter 4 Quiz

1   This is the mean.

3   The range is 92 − 65 = 27.0

5   The mean and the median are equal.

7   The likely low value is 2.0 − 2(0.25) = 1.50 and the likely high value is 2.0 + 2(0.25) = 2.50.

9   Since all of the values are the same, the standard deviation is zero.

Section 5.1
Statistical Literacy and Critical Thinking

1 The word "normal" has a special meaning in statistics. It refers to a specific type of distributions which are characterized by their bell-shaped curve.

3 No. The ten different possible digits are all equally likely, so the graph of the distribution will be flat, not bell-shaped.

5 This statement is sensible.

7 This statement is not sensible. Since the mean is 100 and the distribution is symmetric, 50% of the scores are greater than 100. The number of scores greater than 105 must be less than 50%, not more.

Concepts and Applications

9 Distribution (b) is not normal since it is not symmetric. Distribution (c) has the larger standard deviation since it is more spread out than

11 Normal. It is common for manufactured products, such as coins, to have a distribution that is normal. The weights typically vary above and below the mean weight by the same amounts, so the distribution has one peak and is symmetric.

13 Not normal. The numbers 1 through 49 should occur with approximately equal frequency, so the distribution is uniform, not normal. A graph of the distribution will be flat, not bell-shaped.

15 Normal. Such physical measurements generally tend to be normally distributed. A small number of females will have extremely low counts, a small number will have extremely high counts, and the distribution will tend to peak around the value of the mean.

17 Not normal. The waiting times will tend to be evenly distributed over a 10-minute interval and will therefore have a uniform distribution.

19 The movie lengths are not normally distributed because movies generally have a minimum length, but no maximum length. There are very few movies that are a lot shorter than a typical length, but there are some very long movies.

20 Heart rates are nearly normal because data values are distributed symmetrically about the mean. Many physiological variables have normal distributions.

21 The quarter weights should be nearly normal because the deviations from the mean are symmetrical above and below the mean.

23 a)   The total area under the curve is 1.
    b)   0.20
    c)   0.80
    d)   0.35
    e)   0.45

25 a)   The mean is 115.
    b)   15%
    c)   45%
    d)   15%

Section 5.2
Statistical Literacy and Critical Thinking

1 $z = standard\ score = \dfrac{value - mean}{standard\ deviation} = \dfrac{mean - mean}{standard\ deviation} = 0$

3 No. The distribution of the outcomes from rolling a die is a uniform distribution, not normal. The rule applies only to normal distributions.

5 This statement is not sensible. If we assume that the minimum test score is zero, then the standard deviation cannot be greater than the mean. If it were greater than the mean, at least 16% of the scores would be negative, those that are more than one standard deviation below the mean.

7 The statement does not make sense. A normal distribution does not have two modes.

Concepts and Applications

9 a) 50%. Half of all values are less than the mean.
   b) 84%. 60 is 1 standard deviation to the right of the mean. There is 0.50 to the left of 50 and another 0.34 between 50 and 60.
   c) 2.5%. 70 is 2 standard deviations to the right of the mean. There is 95% between 30 and 70. Half of the remaining 5%, or 2.5%, is to the right of 70.
   d) 84%. 40 is 1 standard deviation to the left of the mean. The percent of the scores between 40 and the mean is 34% and there is another 50% to the right of 50, so there is 84% to the right of 40.
   e) 81.5%. 40 is one standard deviations to the left of the mean. The relative frequency of scores between 40 and the mean is 0.34; 70 is two standard deviations to the right of the mean, so there is 0.475 between 50 and 70. Adding these two frequencies results in 0.815.

11 a) Since 55 seconds is the standard deviation, 68% of the times will be within 55 seconds of the mean of 184.0.
   b) Since 55 seconds is the standard deviation, 95% of the times will be within 110 seconds (2 standard deviations) of the mean of 184.0.
   c) Since 55 seconds is the standard deviation, 99.7% of the times will be within 165 seconds (3 standard deviations) of the mean of 184.0.
   d) Since 294 is 2 standard deviations to the right of 184 (the mean), there is half of 95% or 47.5% between 184 and 294.

13 Since 100 is the mean, z = 0, and 50% of the scores are greater than 100.

15 z = (68 − 100)/16 = −2.00; from Table 5.1, 2.28% of scores are less than 116.

17 z = (132 − 100)/16 = 2.00; from Table 5.1, 97.72% of scores are less than 132.

19 z = (76 − 100)/16 = −1.5; from Table 5.1, 6.68% of scores are less than 76, so 100% - 6.68% = 93.32% are greater than 76.

21 Since 84 and 116 are each 1 standard deviation away from the mean, the a-scores are −1 and +1. From Table 5.1, the area under the curve between these two values is 84.13 − 15.87 = 68.26%.

23 The z-score for 52 is (52 −100)/16 = −3.00. The z-score for 116 is 1.00. The area under the curve to the left of −3.00 is about 0.13%. The area to the left of 1.00 is 84.13%. Then the area between −3.00 and 1.00 is 84.13% - 0.13% = 84.00%.

25 Since the mean is 162 cm, the percentage of heights greater than 162 cm is 50%.

27 z = (156 − 162)/6 = −1.0; from Table 5.1, the percentage less than 156 is 15.87%. The percentage greater than 156 is 100% - 15.87% = 84.13%.

29 z = (177 − 162)/6 = 2.5; from Table 5.1, the percentage less than 177 is 99.38%.

31   z = (144 – 162)/6 = –3.0; from Table 5.1, the percentage less than 144 is 0.13%.

33   The two values of z are z = (156 – 162)/6 = –1.0 and z = (168 – 162)/6 = 1.0. From table 5.1, the percentage between 156 and 168 is 84.13% – 15.87% = 68.26%.

35   The two values of z are z = (148 – 162)/6 = –2.3 and z = (170 – 162)/6 = 1.3. From table 5.1, the percentage between 148 and 170 is 90.32% – 1.07% = 89.25%.

37   In all cases, 5% of coins are rejected. The ranges of weights that are acceptable to the vending machine are

Cent	2.44-2.56 grams
Nickel	4.88-5.12 grams
Dime	2.208-2.328 grams
Quarter	5.530-5.810 grams
Half dollar	11.060-11.620 grams

39   a) $z = \frac{2000-1518}{325} = 1.50$

From Table 5.1, the percentage less than 2000 is 93.32%. Therefore, the percentage higher than 2000 is 100% – 93.32% = 6.68%.

b) $z = \frac{1500-1518}{325} = -0.05$

From Table 5.1, the percentage less than 1500 is 48.01%.

c)   The two standard scores are
$z = \frac{1600-1518}{325} = 0.25 \; and \; z = \frac{2100-1518}{325} = 1.80.$

From Table 5.1, the percentage between 1600 and 2100 is 96.41% – 59.87% = 36.54%.

41   a) $z = \frac{31-30.4}{0.23} = 2.60$

From Table 5.1, the percentage less than 31 is 99.53%. Therefore, the percentage higher than 31 is 100% – 99.53% = 0.47%. [Note that of 50 barometers, this means that only 0.23 barometers read higher than 31, so most of the time, there will be no barometers reading over 31.]

b) $z = \frac{30-30.4}{0.23} = -1.70$

From Table 5.1, the percentage less than 30 is 4.46%.

c)   For the low side, we have (value – 30.4)/0.23 = –1.50. Multiplying both sides of this equation by 0.23, we get value – 30.4 = –0.345. Now adding 30.4 to each side of the equation, we get value = 30.4 – 0.345 = 30.055. For the high side, we have (value – 30.4)/0.23 = 1.50. Multiplying both sides of this equation by 0.23, we get value – 30.4 = 0.345. Now adding 30.4 to each side of the equation, we get value = 30.4 + 0.345 = 30.745. Barometers will be rejected if they read below 30.055 or above 30.745,

d)   The mean of the 50 barometers, 30.4 inches, is the best measure of the actual atmospheric pressure at the time the barometers were read. The distribution of readings could be expected to be approximately normal, so the median could also be used if there were any outliers in the values. Outliers would not affect the median, but may have an effect on the mean.

43 The two standard scores are
$$z = \frac{64-69}{2.8} = -1.80 \text{ and } z = \frac{78-69}{2.8} = 3.20$$

From Table 5.1, the percentage between 64 and 78 is 99.92% − 3.59% = 96.33%. Minitab gives 96.23%.

Section 5.3
Statistical Literacy and Critical Thinking

1 Because the digits are equally likely, they have a uniform distribution. Because the sample means are based on samples of size 3 drawn from a population that does not have a normal distribution, we should not treat the sample means as having a normal distribution.

3 Yes. If the original population is normally distributed, the distribution of the means is normal for all sample sizes, even when the sample size is small.

Concepts and Applications

5 a) The mean of the distribution of sample means is the population mean, 100. With n = 64, the standard deviation of the distribution of the mean is $\sigma/\sqrt{n} = \frac{16}{\sqrt{64}} = \frac{16}{8} = 2$.

   b) With a sample size of n = 100, the mean of the distribution of sample means is the population mean, 100. The standard deviation of the distribution of the mean is $\sigma/\sqrt{n} = \frac{16}{\sqrt{100}} = \frac{16}{10} = 1.6$.

   c) As the sample size n increases, the means tend to be closer together (less variation) and the value of the standard deviation of the mean decreases due to the presence of the square root of n in the denominator.

7 a) The mean of the distribution of sample means is the population mean, 6.5. With n = 81, the standard deviation of the distribution of the sample means is $\frac{\sigma}{\sqrt{n}} = \frac{3.452}{\sqrt{81}} = \frac{3.452}{9} = 0.3836$.

   b) The mean of the distribution of sample means is the population mean, 6.5. With n = 100, the standard deviation of the distribution of the sample means is $\frac{\sigma}{\sqrt{n}} = \frac{3.452}{\sqrt{100}} = \frac{3.452}{10} = 0.3452$.

   c) As the sample size n increases, the means tend to be closer together (less variation) and the value of the standard deviation of the sample mean decreases due to the presence of the square root of n in the denominator.

9 For individual men, we use the population standard deviation, 30, to find the standard score. $z = \frac{value-mean}{standard\ deviation} = \frac{185-170}{30} = 0.5$. From Table 5.1, the percentage with weights less than 185 pounds is 69.15%.
For a sample of size 36, we use the standard deviation of the sample mean, $\frac{\sigma}{\sqrt{n}} = \frac{30}{\sqrt{36}} = \frac{30}{6} = 5.0$, to find the standard score. $z = \frac{value-mean}{standard\ deviation} = \frac{185-170}{5} = 3.00$. From Table 5.1, the percentage of sample means less than 185 pounds is 99.87%.

11 For individual men, we use the population standard deviation, 30, to find the standard scores. $z = \frac{value-mean}{standard\ deviation} = \frac{164-170}{30} = -0.2$; $z = \frac{value-mean}{standard\ deviation} = \frac{176-170}{30} = 0.2$ From Table 5.1, the percentage with weights less than 176 pounds is

57.93% and the percentage with weights less than 164 is 42.07%. Therefore, the percentage with weights between 164 and 176 is 57.93% - 42.07% = 15.86%. For a sample of size 100, we use the standard deviation of the sample mean, $\frac{\sigma}{\sqrt{n}} = \frac{30}{\sqrt{100}} = \frac{30}{10} = 3.0$, to find the standard scores. $z = \frac{value-mean}{standard\ deviation} = \frac{164-170}{3} = -2.00$ and $z = \frac{value-mean}{standard\ deviation} = \frac{176-170}{3} = 2.00$. From Table 5.1, the percentage with mean weights less than 176 pounds is 97.92% and the percentage with mean weights less than 164 is 2.28%. Therefore, the percentage with weights between 164 and 176 is 97.72% - 2.28% = 95.44%.

13   a)   For a sample of size 36, we use the standard deviation of the sample mean, $\frac{\sigma}{\sqrt{n}} = \frac{0.11}{\sqrt{36}} = \frac{0.11}{6} = 0.0183$, to find the standard score. $z = \frac{value-mean}{standard\ deviation} = \frac{12.05-12.00}{0.0183} = 2.70$. From Table 5.1, the percentage of sample means less than 12.05 oz is 99.65%. Thus, the percentage of sample means greater than 12.05 oz is 100% - 99.65% = 0.35%.

   b)   If the true population mean were 12.00 oz, it is very unlikely that we would get a mean amount as high as 12.05 oz for a sample of 36 cans. More than likely, the cans are being filled with more than 12.00 oz. In this case, since the consumers are getting more than they thought they were getting, no one is likely to feel cheated.

15   a)   For individual sizes, we use the population standard deviation, 1.0 inches, to find the standard scores.
   For 6.2, $= \frac{value-mean}{standard\ deviation} = \frac{6.2-6.0}{1.0} = 0.2$. From Table 5.1, the percentage of male head breadths less than 6.2 inches is 57.93%.

   b)   For a sample of size 100, we use the standard deviation of the sample mean, $\frac{\sigma}{\sqrt{n}} = \frac{1.0}{\sqrt{100}} = 0.1$, to find the standard score.
   $z = \frac{value-mean}{standard\ deviation} = \frac{6.2-6.0}{0.1} = 2.00$. From Table 5.1, the percentage of sample means less than 6.2 inches is 97.72%.

   c)   Although the mean head breadth of 100 men is very likely to be less than 6.2 inches, from part (a), we see that only about 58% of men have head breadths less than 6.2 inches. Therefore, about 42% of men have head breadths larger than 6.2 inches and they could not use the helmets.

17   a)   For individual women, we use the population standard deviation, 29 pounds, to find the standard scores.
   For 140 lbs, $z = \frac{value-mean}{standard\ deviation} = \frac{140-143}{29} = -0.10$. From Table 5.1, the percentage with weights less than 140 pounds is 46.02%.
   For 211 lbs, $z = \frac{value-mean}{standard\ deviation} = \frac{211-143}{29} = 2.30$. From Table 5.1, the percentage with weights less than 211 pounds is 98.93%. Thus the percentage with weights between 143 and 211 pounds is 98.93% - 46.02% = 52.91%.

   b)   For a sample of size 36, we use the standard deviation of the sample mean, $\frac{\sigma}{\sqrt{n}} = \frac{29}{\sqrt{36}} = 4.833$, to find the standard scores.
   For 140 lbs, $z = \frac{value-mean}{standard\ deviation} = \frac{140-143}{4.833} = -0.60$.
   For 211 lbs, $z = \frac{value-mean}{standard\ deviation} = \frac{211-143}{4.833} = 14.0$.
   From Table 5.1, the percentage of sample means less than 140 pounds is 27.43%. For 211 pounds, the standard score is off the chart, so at least 99.98% of the weights are less than 211 pounds. Therefore, there is about a 99.98% - 27.43% = 72.55% chance that the mean weight is between 140 and 211 pounds.

   c)   The result from part (a) is more important in redesigning the ejection seats since the seats are occupied by only one person, not a group of 36.

19   a)   For individual quarters, we use the population standard deviation, 0.062 g, to find the standard scores.

For 5.550 g, $z = \frac{value - mean}{standard\ deviation} = \frac{5.550 - 5.670}{0.062} = -1.90$.

From Table 5.1, the percentage with weights less than 5.550 g is 2.87%.

For 5.790 g, $z = \frac{value - mean}{standard\ deviation} = \frac{5.790 - 5.670}{0.062} = 1.90$.

From Table 5.1, the percentage with weights less than 5.790 g is 97.13%. Therefore 2.87% of the quarters will weigh more than 5.790 g. Altogether 5.74% of the quarters will be rejected. If 280 quarters are inserted in the vending machine, 5.74% of them, or 16, can be expected to be rejected.

b)   For a sample of size 280, we use the standard deviation of the sample mean, $\frac{\sigma}{\sqrt{n}} = \frac{0.062}{\sqrt{280}} = 0.0037$, to find the standard scores.

For 5.550 g, $z = \frac{value - mean}{standard\ deviation} = \frac{5.550 - 5.670}{0.0037} = -32.43$.

From Table 5.1, the percentage of sample means less than 5.550 g is less than 0.02%. Similarly, for 5.790 g, z = 32.43, and the percentage of means less than 5.790 is greater than 99.98%. There is at least a 99.96% chance that the mean of 280 quarters is between the limits.

c)   The result from part (a) is more important. No one puts 280 quarters in a vending machine. Individual legitimate quarters rejected may mean the loss of a sale. However, the configuration is designed to prevent losing merchandise when someone puts in a slug or foreign coin about the size of a quarter.

21   a)   For individual students, we use the population standard deviation, 3.1 cm, to find the standard score. $z = \frac{63 - 61.2}{3.1} = 0.58$. From Table 5.1, about 72% of the sitting heights are less than 63 cm. Therefore 28% of the heights are greater than 63 cm.

b)   The standard deviation of the distribution of sample means for n = 25 will be $\frac{\sigma}{\sqrt{n}} = \frac{3.1}{\sqrt{25}} = \frac{3.1}{5} = 0.62$. $z = \frac{63 - 61.2}{0.62} = 2.90$. From Table 5.1, 99.81% of the means will be less than 63 cm. Therefore, about 0.2% of the sample means will be greater than 63 cm.

c)   Part (a) is more relevant since the seats will be occupied by individual girls, not groups of girls.

Chapter 5 Review Exercises

1   a)   The 38 outcomes will occur with equal likelihood. Therefore, they have a uniform distribution, not a normal distribution.

b)   Weights of a homogeneous population, such as 12-year old girls, typically have a normal distribution.

c)   Scores on such tests tend to have a nearly normal distribution since the population is quite homogeneous.

3   a) $z = \frac{value - mean}{standard\ deviation} = \frac{99.0 - 98.20}{0.62} = 1.29$; this corresponds to the 90th percentile.

b)   From part (a), the standard score is 1.29.

c)   No, the data value is less than 2 standard deviations from the mean.

d)   For a sample of size n = 50, the standard deviation of the distribution of means is $\frac{\sigma}{\sqrt{n}} = \frac{0.62}{\sqrt{50}} = 0.0877$. For 97.98, $z = \frac{value - mean}{standard\ deviation} = \frac{97.98 - 98.20}{0.0877} = -2.50$. This corresponds to the 0.62 percentile, so the likelihood that the mean body temperature is 97.98 degrees or lower is

0.0062.

e)   For 101.00, $z = \dfrac{value-mean}{standard\ deviation} = \dfrac{101.00-98.20}{0.62} = 4.52$.   This is an unusual temperature since it is more than two standard deviations above the mean.   We conclude that the person has a fever.

f)   The 95th percentile is associated with a standard score of 1.65.   Thus the temperature must be 1.65 standard deviations above the mean, or 1.65 x 0.62 = 1.02 degrees above 98.20 degrees, or 99.22.   [If you choose to use the closest table value, use z = 1.6.   Then the temperature is 98.20 + 0.99 = 99.19, not very different.]

g)   The 5th percentile is associated with a standard score of -1.65.   Thus the temperature must be 1.65 standard deviations below the mean, or 1.02 degrees below 98.20 degrees, or 97.18.   [Again, if you choose to use -1.60, the temperature will be 0.99 degrees below 98.20, or 97.21.]

h)   For 100.6, $z = \dfrac{value-mean}{standard\ deviation} = \dfrac{100.60-98.20}{0.62} = 3.87$.   This corresponds to a percentile higher than 99.98.   Thus fewer than 0.02% of healthy adults would be expected to have a temperature above 100.6.   It should be very safe to conclude that someone with a temperature of 100.6 or higher has a fever.   On the other hand, using such a high cutoff point will certainly result in concluding that a number of people do not have fevers when, in fact, they do have a fever.

i)   For a sample of n = 106, the standard deviation of the distribution of sample means is $\dfrac{\sigma}{\sqrt{n}} = \dfrac{0.62}{\sqrt{106}} = 0.0602$.   For a temperature of 98.2, $z = \dfrac{value-mean}{standard\ deviation} = \dfrac{98.20-98.60}{0.0602} = -6.64$.   Thus the sample mean is about 6.6 standard deviations below the mean; the chance of selecting such a sample is extremely small if the assumed mean is correct.   The assumed mean (98.60) may be incorrect.

Chapter 5 Quiz

1   Statements (b),(c), and (d) are correct.   Statement (a) is not correct because there are many distributions that are not unusual that are not normal (e.g., uniform).   Statement (e) is not correct because if the standard deviation were zero, all of the values would be the same.   The distribution would not be bell-shaped.

3   Eight hundred pounds either side of 2400 pounds is two standard deviations. For z = -2 and z = 2, Table 5.1 gives 2.28% and 97.72% respectively, so the percentage between 1600 and 3200 is 97.72% - 2.28% = 95.44%.

5   The standard deviation of the sample means is $\dfrac{\sigma}{\sqrt{n}} = \dfrac{49}{\sqrt{100}} = 4.9$.

7   From the previous exercise, z = -1.00.   From Table 5.1, 15.87% of the scores are less than 801, so 100% - 15.87% = 84.13% of the scores are greater than 801.

9   Since 801 is 1 standard deviation below the mean of 850 and 948 is 2 standard deviations above the mean of 850, From Table 5.1, the percentage of scores between 801 and 948 is 97.72% - 15.87% = 81.85%.

Section 6.1
Statistical Literacy and Critical Thinking

1 No, the argument is not valid. With 1000 births, any specific number of girls will have a very low probability, but 501 girls is very close to the 500 girls that is expected, so the result is not evidence that the method is effective.

3 No. Statistical significance at the 0.05 level means that there is less than a 0.05 probability that the result occurred by chance, but a probability less than 0.05 is not necessarily less than 0.01.

5 This statement is not sensible. Instead of referring to a meaningful effect, statistical significance refers to an observed result that is unlikely to occur by chance. The unemployment rate may have practical significance, but we cannot make a statement about statistical significance without reference to any actual data.

7 This statement is sensible. With a probability of only 1 chance in a

Concepts and Applications

9 Statistically significant. This result is very unlikely to have occurred by chance.

11 Not statistically significant. With 6 equally likely possible outcomes, we expect that about 1 of the 6 outcomes would be a 6, so not getting any 6s is very close to what we expect and is not unusual.

13 Statistically significant. It is very unlikely that when 20 adults are randomly selected, they are all women (about 1 chance in 1 million).

15 Not statistically significant. This result could easily occur by chance.

17 Significant. One would not expect a difference this great among thirty identical cars, assuming that they were driven under identical conditions.

19 With 945 babies, the number of girls would usually be about half of that (around 472.5), so the result of 879 girls represents a very large departure from what is expected by chance. The result is statistically significant.

21 a)   If 100 samples were selected, the mean temperature would be 98.20 or less in 5 or fewer of the samples if the true mean were 98.60 degrees.

    b)   Selecting a sample with a mean this small is extremely unlikely if the true mean is 98.60 degrees and would not be expected by chance.

23 This result is not significant because the probability of it occurring by chance when there is no real improvement is greater than 0.05.

Section 6.2
Statistical Literacy and Critical Thinking

1 P(A) represents the probability that event A occurs. P(not A) [(or P($\bar{A}$)] is the probability that event A does not occur. For the True/False question, all three probabilities are 0.5 if the student is just guessing.

3 There is one chance in 524,288 that all 20 babies are of the same gender. Such an event is unusual because its probability is so small.

5 This statement makes sense. The probability of a certain event is 1.

7 This statement makes sense. Since either it rains or it does not, the probabilities of these two outcomes must total 1. If the probability of rain is 0.9, then the probability of no rain must be 0.1.

9 This statement does not make sense. If this were the case, half of the people in the world would be struck by lightning each year, but clearly that does not happen. Jack's estimate of 0.5 is much too high to be reasonable.

Concepts and Applications

11   There are six possible outcomes, four of which are greater than 2. Assuming that the die is fair (the outcomes are equally likely), the probability of a result greater than 2 is 4/6 or 2/3.

13   Assuming that the roulette wheel is fair, there are 38 equally likely outcomes and 18 of them are red, so the probability is 18/38 or 0.47.

15   Assuming that the 365 days of the year are equally likely, there is 1 chance out of 365 that the birthday is the same as yours.

17   Assuming that girls and boys are equally likely, the probability is 1/2. The fact that their first two children were boys makes no difference.

19   When you roll one die, you must get an outcome less than 7. The probability of getting an outcome not less than 7 is therefore zero.

21   P(Monday) = 1/7, so P(not Monday) = 1 − 1/7 = 6/7.

23   P(miss) = 1 − P(make) = 1 − 0.55 = 0.45.

25   The probability of getting a hit is 0.280, so the probability of not getting a hit is 1 − 0.280 = 0.720.

27   Altogether, there are 45 M&Ms. Assuming that each is equally likely to be picked, P(Red) = Number of Red/Total Number = 10/45 = 2/9. Similarly, P(Blue) = 15/45 = 1/3, and P(Yellow) = 20/45 = 4/9. P(Not Yellow) = 1 − P(Yellow) = 1 − 4/9 = 5/9.

29   There are eight possible outcomes: GGG, BGG, GBG, GGB, BBG, BGB, GBB, BBB
For each of the parts below, the solution is obtained by dividing the number of outcomes in the event by 8.
a)   1/8 = 0.125   (GGG)
b)   3/8 = 0.375   (BBG, BGB, GBB)
c)   1/8 = 0.125   (GBB)
d)   7/8 = 0.875   (GGG, BGG, GBG, GGB, BBG, BGB, GBB)
e)   4/8 = 0.5   (BBG, BGB, GBB, BBB)

31   The forecaster has been right 18 out 30 times, a relative frequency of 18/30 = 0.60. Thus, his probability of being correct on the next forecast is 0.6.

33   P(Success) = 0.86.

35   The probability that a person you meet at random will be over 65 is 34.7/281 = 0.123 in 2000 and will be 78.9/394 = 0.200 in 2050. Thus your chances will be greater in 2050.

37 a)

Outcomes for Tossing Four Fair Coins					
Coin 1	Coin 2	Coin 3	Coin 4	Outcome	Probability
H	H	H	H	HHHH	1/16
H	H	H	T	HHHT	1/16
H	H	T	H	HHTH	1/16
H	H	T	T	HHTT	1/16
H	T	H	H	HTHH	1/16
H	T	H	T	HTHT	1/16
H	T	T	H	HTTH	1/16
H	T	T	T	HTTT	1/16
T	H	H	H	THHH	1/16
T	H	H	T	THHT	1/16
T	H	T	H	THTH	1/16
T	H	T	T	THTT	1/16
T	T	H	H	TTHH	1/16
T	T	H	T	TTHT	1/16
T	T	T	H	TTTH	1/16
T	T	T	T	TTTT	1/16

b)

Probability Distribution for the Number of Heads in Tossing Four Fair Coins	
Result	Probability
4 Heads (0 Tails)	1/16
3 Heads (1 Tail)	4/16
2 Heads (2 Tails)	6/16
1 Head (3 Tails)	4/16
0 Heads (4 Tails)	1/16
Total	**1**

c)   6/16 = 0.375
d)   15/16 = 0.9375
e)   2 heads (Probability is 0.375)
d)   The deviations are what might be expected by chance.

Section 6.3
Statistical Literacy and Critical Thinking

1 If some process is repeated through many independent trials, the proportion of trials in which some event occurs will be close to the probability of the event. As the number of trials increases, the proportion of trials in which the event occurs will get closer to the probability of the event.

3 His betting strategy is unwise because the casino continues to have an advantage. His past performance does not affect future events. His reasoning is not correct. Although his proportion of winning bets is likely to increase, it could increase while he continues to lose more money. [For example, suppose that in 10 coin tosses, heads occurs four times, a 0.40 rate, and the number of heads is only one lower than the 5 expected. The coin tossing continues and after 100 tosses, there are 44 heads. The proportion of heads has increased to 0.44, but the number of heads is now 6 lower than the 50 expected.] This flawed reasoning has caused many gamblers to lose large amounts of money.

5 The statement makes sense. By betting repeatedly on one number on the roulette wheel, he will lose an average of $.26 per play. On the craps bet, he will lose only $.07 per play, so it is better to play the craps bet. On average, Steve will still lose his money, but he will get to play longer.

7 The statement does not make sense since this combination of numbers has the same chance of winning as any other combination.

Concepts and Applications

9 No, you should not expect to get exactly 250 girls since the probability of that particular outcome is extremely small. The proportion of girls should approach 0.5 as the number of births increases.

11 For each of the possible outcomes 1, 2, 3, 4, 5, and 6, your net winnings would be (after subtracting the $10 it costs you to play) –$8, –$6, –$4, –$2, $0, and $2. Each of these outcomes has probability 1/6. Therefore, your expected net winnings =
(–8)(1/6) + (–6)(1/6) + (–4)(1/6) + (–2)(1/6) + (0)(1/6) + (2)(1/6) = –3.
Since your expected net winnings are negative, you should not play.

13 For the 1-point attempt, Expected value = (1 x 0.94) + (0 x 0.06) = 0.94
For the 2-point attempt, Expected value = (2 x 0.37) + (0 x 0.63) = 0.74
The 1-point attempt makes more sense in most cases. However, if a team is two points behind with little or no time left, it makes no sense to go for 1 point. The same is true if the team is behind by 16 points and it is unlikely that the team will have enough time to score three times.

15 Your waiting time is uniformly distributed between 0 and 30 minutes. The center of this symmetric distribution is 15 minutes, so 15 minutes is your expected waiting time.

17 On one ticket, you will spend one dollar and you may get something back. The expected value is –$1 + ($3,000,000 x 1/76,275,360) + ($150,000 x 1/2,179,296) + ($5,000 x 1/339,002) + ($150 x 1/9,686) + ($100 x 1/7,705) + ($5 x 1/220) + ($5 x 1/538) + ($2 x 1/102) + ($1 x 1/62) = –$0.78. If you spend $365 per year on the lottery, you can expect to lose 365 x $0.78 = $284.70. If you find the expected value with greater precision, you get $285.02.

19 Your expected value is (–$5)(251/495) + ($5)(244/495) = –$0.0707. Since this is for a $5 bet, you lose 1/5 of this amount for each dollar bet, or about 1.4 cents.

21 a) **Decision 1** – Option A: Expected value = $1,000,000
Option B: Expected value = (%2,500,000 x 0.10) + ($1,000,000 x 0.89) + ($0 x 0.01) = $1,140,000
Decision 2 – Option A: Expected value = ($1,000,000 x 0.11) + ($0 x 0.89) = $110,000

Option B: Expected value = ($2,500,000 x 0.10) +
($0 x 0.90) = $250,000

b)   Responses are not consistent with expected values in Decision 1, but they are consistent in Decision 2.

c)   It appears that people choose the certain outcome ($1,000,000) in Decision 1.

23   a)   Three of the six numbers are even, so the probability is 0.5; three of the six numbers are odd, so the probability is 0.5.

b)   You have gained $45 and lost $55, so you have a net loss of $10.

c)   You have gained $92 and lost $108, so you have a net loss of $16.

d)   You have gained $240 and lost $260, so you have a net loss of $20.

e)   After 100 tosses, it was 45%; after 200 tosses, it was 92/200 or 46%; and after 500 tosses, it was 240/500 or 48%. This illustrates the gambler's fallacy because, even though the percentage of wins is increasing, the amount of the net loss is also increasing.

f)   You would need to have 300 heads in 600 tosses. Since you already have 240 heads in the first 500 tosses, you would need 60 more heads in the last 100 tosses.

Section 6.4
Statistical Literacy and Critical Thinking

1   Because they are expressed as the numbers of births for each 1000 people in the population, the birth rates can be compared directly. The actual numbers of births should be considered in the context of the population sizes of the countries, so a comparison would require the numbers of births along with population sizes, and even then the comparison would not be easy because of the numbers involved.

3   Life expectancy is the number of years a person with a given age can expect to live on average. A 30-year-old person will have a shorter life expectancy than a 20-year-old person. The 30-year-old person is not expected to live as many additional years as a 20-year-old person.

5   The statement does not make sense. One cannot live a shorter life than he or she has already lived.

7   This statement makes sense. An older person has already survived a period during which others of the same age have died. Because this person has already survived a period during which others died, this older person is expected to live to a greater age.

9   For 2000, the fatality rate per thousand departures was 92/9035 = 0.0102; For 2004, the fatality rate per thousand departures was 14/11182 = 0.0013; For 2008, the fatality rate per thousand departures was 3/10437 = 0.000287. The year 2008 was the safest because it had the lowest number of fatalities per 1000 departures.

11   For 2000, the fatality rate per million passengers was 92/666.2 = .1381; For 2004, the fatality rate per million passengers was 14/697.8 = .0201; For 2008, the fatality rate per million passengers was 3/690.2 = .0043. The year 2008 was the safest because it had the lowest number of fatalities per million passengers.

13   This answer is found in the fourth column and the 19-20 row: 58.8 years.

15   This answer is found in the second column and the 19-20 row, which gives the probability of dying in the next year as 0.000833. Multiply this by 10,000 to find the death rate per 10,000 people as 8.33.

17   For Utah, there were (55,633/2,736,424) x 1000 = 20.33 births per 1000 population. For Maine, there were (13,610/1,316,456) x 1000 = 10.34 births per 1000 population.

19   a)   Note that 313 million is the same as 313,000 thousand. If the birth rate is 13.8 births per 1,000, then the total number of births is approximately 13.8 x 313,000 = 4,319,400.

b) The total number of deaths is approximately 8.4 x 313,000 = 2,629,200.

c) In 2011, the U.S. population rose approximately 4,319,400 – 2,629,200 = 1,690,200 taking into account only births and deaths.

d) The rate of population growth in 2011 was 1,690,200/313,000,000 = 0.0054 or 0.54%.

Section 6.5
Statistical Literacy and Critical Thinking

1 The two events are independent since the occurrence (or non-occurrence) of turning on either device and finding that it works does not change the probability of the other device and finding that it works.

3 If all students are available for the second selection, this is sampling with replacement. The second outcome is independent of the first.

5 This statement is not sensible. The lottery numbers are selected in a way that is independent of any previous results, so the outcomes of previous lottery selections have no effect on future results.

7 This statement is sensible. If the probability of event *A* is 0.5, the probability of events *A* or *B* must be at least 0.5 and so it could be 0.8.

9 Since the births are independent, P(Fourth child is a girl) = 0.5.

11 Since the selections are independent, P(A and 1 and 2 and 3 and 4) = P(A)P(1)P(2)P(3)P(4) = (1/26)(1/10)(1/10)(1/10)(1/10) = 1/260,000.

13 Since the selections can be repeated, the probabilities of each type remain the same for each selection: 30/60 = 1/2 for rock, 15/60 = 1/4 for jazz, and 15/60 = 1/4 for blues.

a) P(four jazz selections in a row) = 1/4 x 1/4 x 1/4 x 1/4 = 1/256 = 0.0039.

b) P(five blues in a row) = 1/4 x 1/4 x 1/4 x 1/4 x 1/4 = 1/1024 = 0.00098.

c) P(jazz and then rock) = 1/4 x 1/2 = 1/8 = 0.125.

d) Since P(non-rock selection) = 1 – P(rock) = 1 – 1/2 = 1/2, P(four non-rock selections in a row) = ½ x ½ x ½ x ½ = 1/16 = 0.0625

e) There are 60 equally likely songs available for each selection. No matter which song is played first, the probability that the next one is the same is 1/60 = 0.0167.

15 The number of people who either pled guilty or were sent to prison is 392 + 564 + 58 = 1014. Therefore P(guilty plea or sent to prison) = 1014/1028 = 0.986.

17 Altogether, 956 pled guilty out of the total of 1028.
P(First pled Guilty and Second pled Guilty) = P(First pled Guilty) x P(Second pled Guilty *given* First pled Guilty) = (956/1028)(955/1027) = 912980/1055756 = 0.865.

19 Of the total of 1028 defendants, 392 pled guilty and were sent to prison, so P(Pled Guilty and Went to Prison) = 392/1028 = 0.381.

21 The number of accidents in which the pedestrian was intoxicated or the driver was intoxicated is 59 + 79 + 266 = 404. Therefore P(the pedestrian was intoxicated or the driver was intoxicated) = 404/985 = 0.410.

23 The number of accidents in which the pedestrian was intoxicated or the driver was not intoxicated is 59 + 266 + 581 = 906. Therefore P(the pedestrian was intoxicated or the driver was not intoxicated) = 906/985 = 0.920.

25 Of the 985 accidents, 138 involved intoxicated drivers. When two accidents are selected without replacement, P(First Driver intoxicated and Second driver intoxicated) = P(First Driver intoxicated) x P(Second driver intoxicated *given* First Driver intoxicated) = (138/985)(137/984) = 0.0195.

27 a) P(drug or placebo) = (120 + 100)/300 = 220/300 = 0.733

b) P(improved or not improved) = (138 + 162)/300 = 1

c)   We can work this one in two ways: P(drug or improved) = (65 + 55 + 42 + 31)/300 = 193/300 = 0.643; also P(drug or improved) = P(drug) + P(improved) – P(drug and improved) = 120/300 + 138/300 – 65/300 = 193/300 = 0.643.

d)   P(drug and improved) = 65/300 = 0.22.

29 a)   Since the events are independent, P(A from father and A from mother) = P(A from father) x P(A from mother) = 0.75 x 0.75 = 0.5625.

b)   P(Aa or aA) = P(Aa) + P(aA) = P(A) x P(a) + P(a) x P(A) = 0.75 x 0.25 + 0.25 x 0.75 = 0.375.

c)   P(aa) = P(a) x P(a) = 0.25 x 0.25 = 0.0625

d)

Event	Probability
AA	0.5625
Aa	0.1875
aA	0.1875
aa	0.0625

e)   P(Recessive trait) = P(aa) = 0.0625.   P(Dominant trait) = 1 – 0.0625 = 0.9375

31 a)   P(6) = p = 1/6.   P(he wins) = P(at least one 6 in four trials) = 1 – P(no sixes in 4 trials) = $1 - (1 - 1/6)^4 = 1 - (5/6)^4$ = 1 – 625/1296 = 671/1296 = 0.518, so if played repeatedly, he should expect to win more often than he loses.

b)   P(double 6) = p = 1/36.   P(he wins) = P(at least one Double 6 in 24 trials) = 1 – P(no double sixes in 24 trials) = $1 - (1 - 1/36)^{24} = 1 - (35/36)^{24}$ = 1 – 0.509 = 0.491, so if played repeatedly, he should expect to lose more often than he wins.

Chapter 6 Review Exercises

	Head Injuries	No Injury	Total
Wore Helmet	96	656	752
No Helmet	480	2330	2810
Total	576	2986	3562

1   P(Head Injury) = (576)/3562 = 0.162

3   P(No helmet or Not Injured) = P(No Helmet) + P(Not Injured) – P(No helmet and not injured) = 2810/3562 + 2986/3562 – 2330/3562 = 3466/3562 = 0.973

5   P(No Helmet and not injured) = 2330/3562 = 0.654

7   P(No helmet given head injuries) = 480/576 = 0.8333

9   a)   P(Not Good) = 1 – P(Good) = 1 – 0.27 = 0.73

b)   P(Both good) = P(Good1 and Good2) = P(Good1)P(Good2) =(0.27)(0.27) = 0.0729.

c)   Expected number of Good chips = 0.27 x 5 = 1.35.

d)   P(All 5 good) = 0.27^5 = 0.001435. Getting 5 good ones in 5 selections has a very small probability if 27% of the chips are good, so we would tend to believe that the true yield is greater than 27%.

Chapter 6 Quiz

1 P(Correct) = 0.7, so P(Wrong) = 1 − 0.7 = 0.3.

3 Answers may vary. This doesn't happen very often, so an estimate of 0.01 or lower is reasonable.

5 $P(\bar{A}) = 1 - P(A) = 1 - 0.4 = 0.6$

7 P(Group B or Passed) = (562 + 10)/586 = 0.976.

9 P(Group A and Passed) = 10/586 = 0.0171

<u>**Section 7.1**</u>
Statistical Literacy and Critical Thinking

1 r is called the linear correlation coefficient and it measures how well a set of paired data values fit the pattern of a straight line.

3 No. It is possible that there is a relationship between the two variables that is not linear, that is, it does not follow a straight line pattern.

5 This statement is not sensible. There might be some other factor that caused the stork population to increase and the births to increase at the same time. One such factor might be a growing population that resulted in an increase in the number of births and an increase in the number of housing structures that provide good nesting locations for the storks. It is not likely that there is a direct cause-effect relationship between the two variables.

7 This statement is not sensible. The strength of a correlation is a measure of how close the data points fall to a straight line, not of how large the sample is. A large sample could have zero correlation while a small sample could have a correlation near 1 or -1.

Concepts and Applications

9 These variables will be positively correlated because apples are typically sold by the pound, so a heavier bag will have a greater cost.

11 These variables will be negatively correlated because heavier planes tend to get lower miles per gallon of fuel.

13 These variables will be uncorrelated because points in a Super Bowl game have no relationship at all to the DJIA.

15 These variables are not correlated.

17 There is a strong positive correlation, with the correlation coefficient approximately 0.8 to 0.9.

19 a)

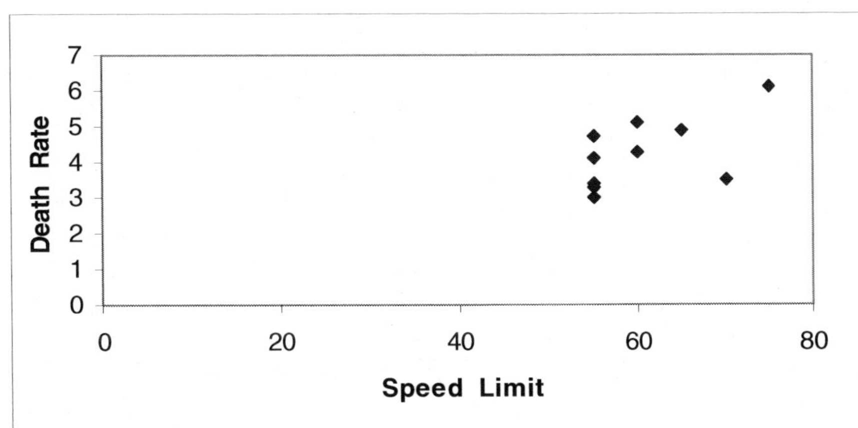

b) There is a moderate positive correlation; r = 0.59 exactly.

c) It is difficult to argue with the phrase "no guarantee". However, four of the five lowest death rates are associated with a 55 mph speed limit. With the exception of Britain, the higher speed limits are generally associated with higher death rates. Death rates are also influenced by other factors beside speed limits.

See the end of this chapter's solutions for the computation details for r for this exercise.

21   a)

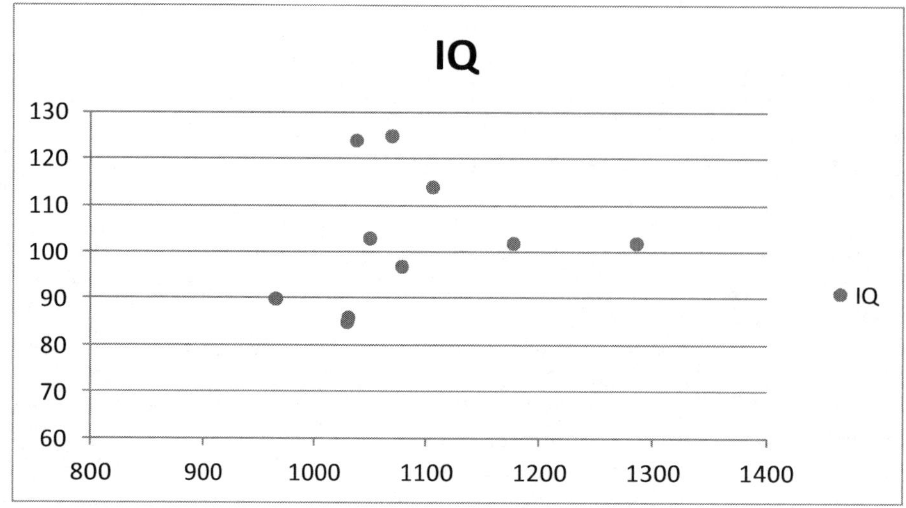

b)   There is a slight positive correlation (r = 0.179).
c)   No. The correlation is very low and based on only a few data points.
The data do not appear to have a linear relationship.

23   a)

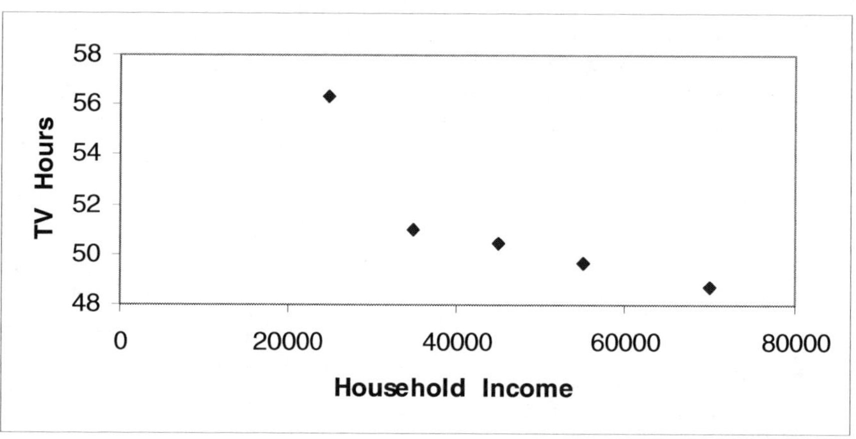

b)   There is a strong negative correlation between income and the number of
TV hours per week (r = -0.86 exactly).
c)   Families with more income have more opportunities to do other things.
It may also be that some people have higher incomes because they spend
more time associated with earning money, leaving less time to watch TV.
Obviously, just watching less TV will not increase your income, but if
you use that time for income producing activities, your income could
increase.

25   a)

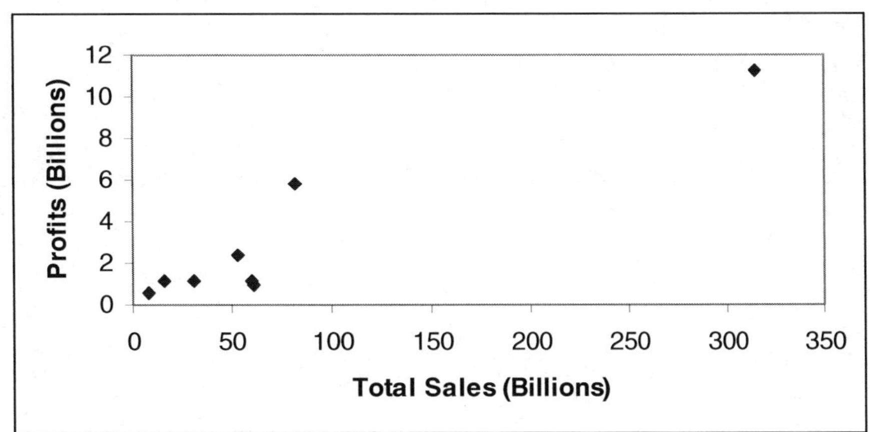

b)   There is a strong correlation between sales and earnings (r = 0.94 exactly). The strong correlation in this case is highly affected by the Wal-Mart data, although the general trend for the rest of the data still shows a positive correlation.

c)   Higher sales do not necessarily translate into higher earnings. Some companies may have larger expenses, driving earnings down.

27   True. Since correlation measures how closely the points are to a straight line, the correlation should be the same if the variables are interchanged. Also, interchanging x and y in the formula for r results in exactly the same formula.

Section 7.2
Statistical Literacy and Critical Thinking

1   The correlation is consistent with the possibility that Lisinopril lowers blood pressure, but without further study we cannot be sure that this is the case since the correlation could be due to other factors or coincidence.

3   Outliers are values that are very far away from almost all of the other values in a data set, so the $1 salary is an outlier. Whatever the second variable is, outliers might make it appear that there is a significant correlation that is not real, or they might mask a real correlation.

5   The statement is not sensible. We can conclude that there is no linear relationship, but there might be some other curved type of relationship.

7   The statement is sensible.

Concepts and Applications

9   There is a positive correlation between the number of unregistered handguns and the crime rate. This correlation is probably due to a common underlying cause. Many crimes are committed with unregistered handguns, some with no handguns. It is also possible that a rising crime rate leads people to purchase handguns for protection.

11   There is a positive correlation that is due to a direct cause. As students study more, they gain a better understanding of the subject and their test scores are likely to be higher.

13   This is a positive correlation that is probably due to a common cause, such as the general increase in the number of cars and traffic. That is, the increase in vehicles requires that some intersections get new lights to control the traffic and, at the same time, the increase in vehicles is in part responsible for the increased number of crashes.

15   There is a negative correlation that is probably due to a direct cause. As gas prices increase, people can't afford to drive as much or as far, so they cut costs by driving less.

17   a)   The point (0.4, 1.0) is an outlier since it lies far from the rest of the data points. Without the outlier, there is no linear relationship between the variables, so the correlation is zero.

      b)   With the outlier included, there is a negative correlation (actual value is r = -0.57).

19   a)

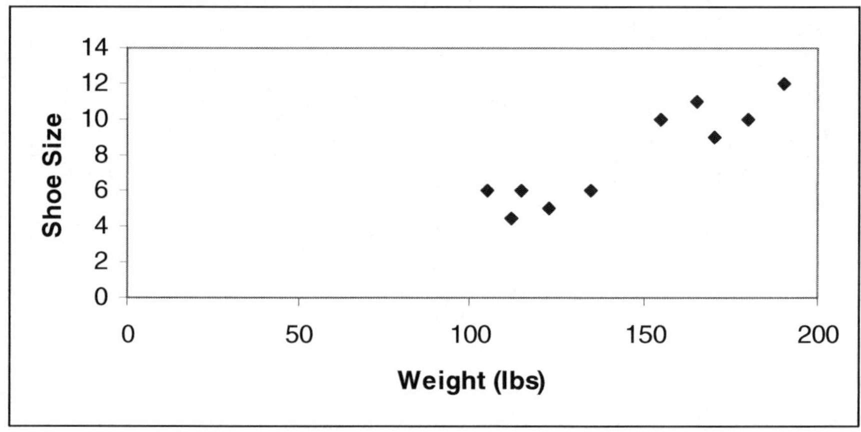

The actual correlation coefficient is r = 0.92, so there is a very strong correlation between weight and shoe size.

      b)   If you look at just the first five points, there does not appear to be a strong linear relationship between weight and shoe size. The same is true if you look only at the last five points. It appears that the apparent correlation found in part (a) is more a result of gender than of weight.

21   a)   The actual correlation coefficient is r = 0.77. This is significant at the 0.01 level, indicating a strong correlation.

      b)   The 16 points to the left correspond to relatively affluent countries, such as Sweden, with low birth rates and low death rates. The remaining points on the right correspond to relatively poor countries, such as Uganda, with higher birth rates and higher death rates.

      c)   There appears to be a negative correlation between the variables for the wealthier countries and a positive correlation for the poorer countries. You could confirm the correlations by dividing the data into two groups and finding the correlation in each group. You can confirm the location of the data points for individual countries such as Sweden and Uganda by going to the internet and entering "birth rates" in Google and choosing one of the web sites that shows birth and death rates for different countries. One such site is http://www.os-connect.com/pop/p3n.asp. There you will find that the birth and death rates as of Jan. 2, 2012 for Norway are 10.84 and 9.24, while for Uganda, they are 47.49 and 11.71. A scatter diagram for all 199 countries shown on this site is shown below. The correlation coefficient for all the data is r = 0.47.

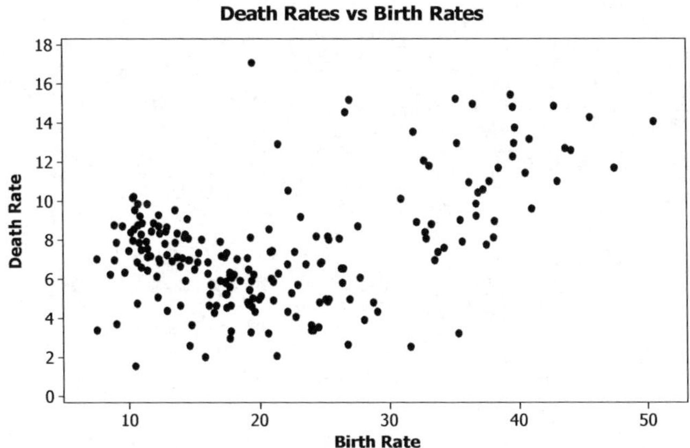

<u>**Section 7.3**</u>
Statistical Literacy and Critical Thinking

1   A best-fit line (or regression line) is a line on a scatter diagram that lies closer to the data points than any other possible line (according to the standard statistical measure of closeness). That is, if you find the difference between each data y value and the corresponding y value for the point directly above or below it on the line, square that difference, and then total the squares of the differences, the best-fit line has the lowest total possible. A best-fit line is useful for predicting the value of the y variable given some value of the other variable (x variable) within the range of the x values in the data set.

3   She should use a multiple regression model. In linear regression, there is only one independent (x) variable. In multiple regression, there are two or more independent variables. In this case, the investigator wants to find the best-fitting equation that gives the heights of the daughters in terms of the heights of the mothers and fathers.

5   This statement makes sense since the total cost of the gas is a constant multiple of the number of gallons purchased (if all of the gas is purchased at the same price).

7   This statement does not make sense. A woman 120 inches tall would be 10 feet tall, but that is well beyond the scope of the data, and it does not make sense to make predictions outside the scope of the data.

Concepts and Applications

9   a)   Here is one way to add a best-fit line to the scatter diagram. Enter the Color and Price data from Table 7.1 into an Excel 2007 spreadsheet with Color in the left hand column. Highlight the two sets of data (including column headings) and click on the **Insert** tab icon at the top of the spreadsheet. Select the **Scatter** icon option and use the top left sub-type option. The graph should appear on the spreadsheet. Now right click on any one of the data points in the graph and click on **Add Trendline**. Select the **Linear** option and click on **Close**. If you want to add axis labels, click on the **Text box** icon and then click on the axis where you want the text box to go. We have added the label "Color" to the horizontal axis in the plot below.

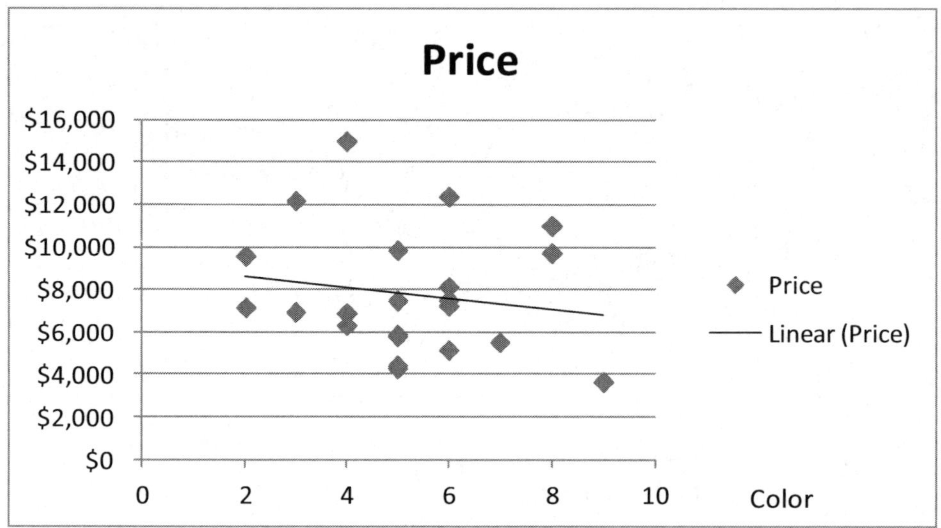

b)   You can use Excel to find the correlation as well. Assuming that you
        have placed the data in columns A and B, rows 2 though 24, in any new
        cell enter the expression **=correl(a2:a24,b2:b24)** and press Enter. The
        actual r value will appear as -0.16323; squaring this, we find that r^2
        = 0.026645, so about 3% of the variation in price can be explained by
        the best-fit line.

c)   The best-fit line should not be used to make predictions since r^2 is so
        close to zero.

**In Exercises 10-12, your solution may differ slightly from those shown as a
result of estimating the data values from graphs.**

11   a)

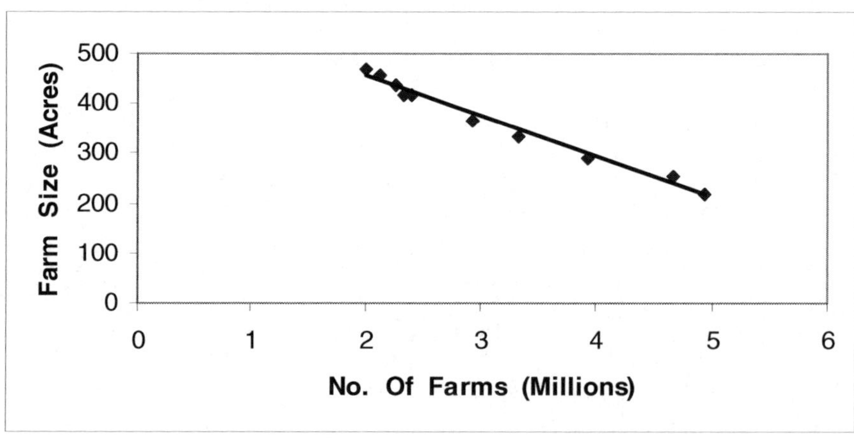

b)   Actual r = -0.99;
        r^2 = 0.97. About 97% of the variation in farm size can be explained by
        the best-fit line.

c)   The best-fit line could be used to make predictions within the range of
        the number of farms included in the data. Because there does appear to
        be a slight curvature to the points, predicting outside that range
        should not be done.

In Exercises 13-20, we will show the actual best-fit lines in part (a). Your line may be different since you will have tried to draw the line by eye. Similarly, in part (b), we will show the actual values of r^2. Your estimates may be different.

13   a)

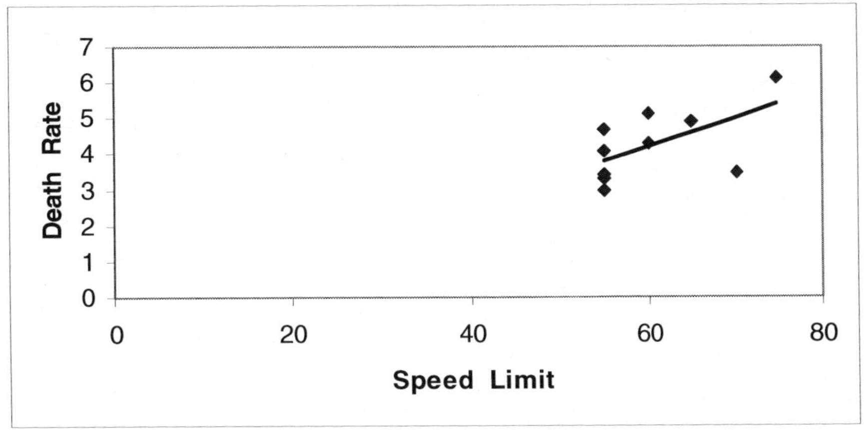

b)   The correlation is moderately positive. The actual value of r is 0.587, so r^2 = 0.345. Thus, about 35% of the variation in the death rate can be accounted for by the linear relationship with the speed limit.

c)   (75,6.1) and (70,3.5) are both possible outliers, the first because it is away from most of the data points and the latter because the death rate is lower than might be expected considering the rest of the data. Because one point is above the best-fit line and one is below, the net effect of the two points is probably to cancel each other out since both points will "pull" the line toward themselves.

d)   No. It would be difficult to conclude that the model is necessarily linear outside the range of the data. Higher speed limits might cause the death rate to increase faster than indicated by the line while lower speed limits might cause the death rate to decrease faster than indicated by the line. Furthermore, the value of r is too small to consider predictions based on the best-fit line to be reliable.

See the end of this chapter's solutions for the details of the computation of the best-fit line for this exercise.

15   a)

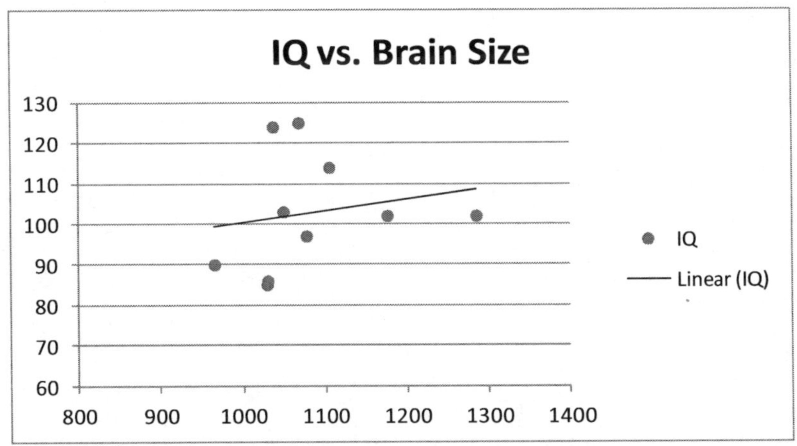

b)   The correlation between IQ and brain size is very weak. The actual value of r is 0.179; r^2 = 0.032. Only about 3% of the variation in IQ can be accounted for by a linear relationship with brain size.

c)   There is one possible outlier at (1285, 102). The effect of this outlier is to pull the right end of the line down slightly.

d)   This best-fit line is worthless for predicting IQ from brain size. It definitely should not be used for prediction of IQ based on brain size.

17   a)

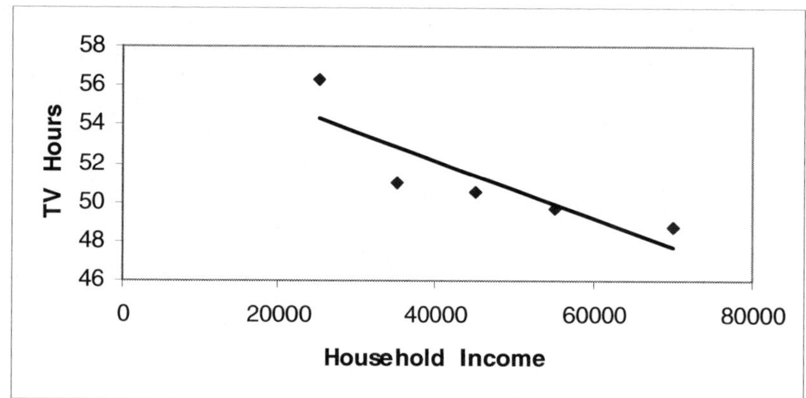

b)   The correlation is fairly strong. The actual value of r is −0.86 and r^2 = 0.74. About 74% of the variation in TV hours can be attributed to the linear relationship with household income.

c)   The data point in the upper left of the diagram is a possible outlier since it is clearly out of line with the other four data points. The effect of this point is to pull the left end of the best-fit line upwards. With this point removed, the correlation would be −0.997.

d)   No. Predicting outside the range of the data points is seldom a good idea. Even though the correlation is fairly strong for these data, remember that the two endpoints are based on assumptions that $25,000 and $70,000 can adequately represent the two end income classes. Having only five data points, one of which is an outlier, is another good reason for not using these data for predictions based on the best-fit line.

19   a)

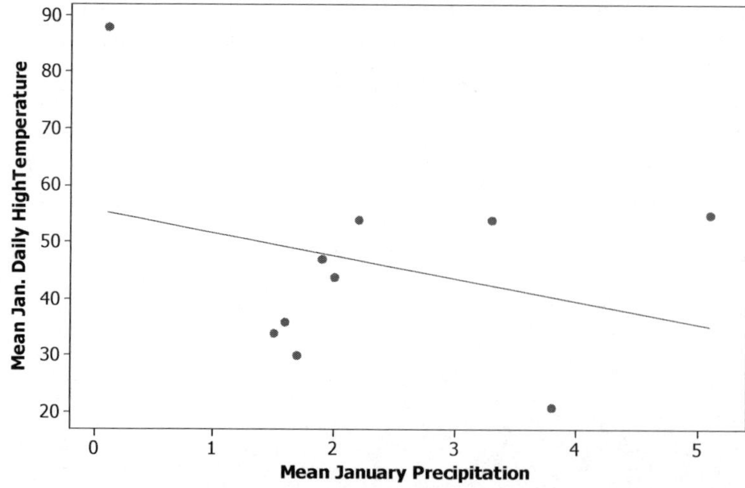

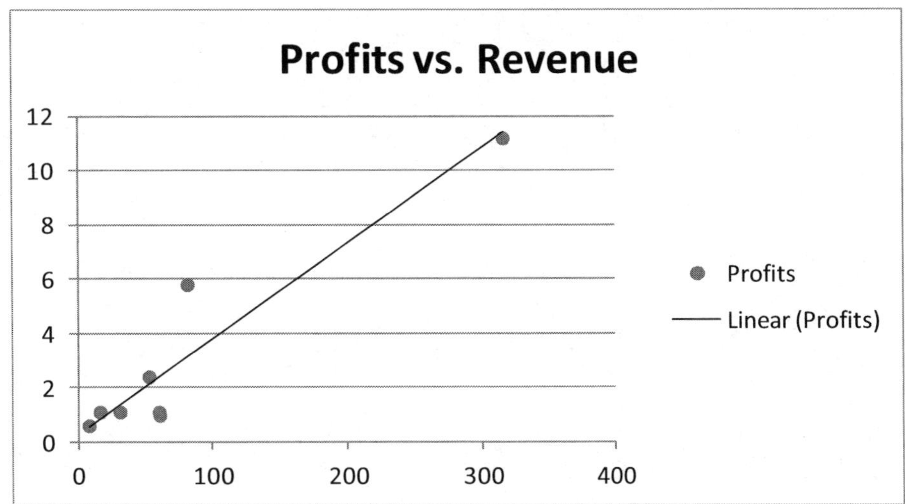

b)   The correlation is very strong. The actual value of r is 0.941 and
     r^2 = 0.886. About 89% of the variation in profits can be accounted for
     by a linear relationship with total sales.
c)   The data point (315.6, 11.2) in the upper right of the diagram for
     Walmart is clearly an outlier, lying far away from the rest of the data
     points. Because that point is close to being in line with the other
     points, deleting it does not change the line very much. However, the
     value of r drops to about 0.69.
d)   Due to the lack of data between Walmart and the other points, the
     linear model should not be used for predictions either within or
     outside the range of data values. In fact, there may not be any data
     points possible to the right of Walmart.

Section 7.4

Statistical Literacy and Critical Thinking

1   A correlation between two variables (1) could be the result of as
     coincidence; (2) could be due to a common cause; or (3) could be due to a
     direct influence of one of the variables on the other.
3   A confounding variable is a variable that is not included in the analysis,
     but it affects the variables that included in the analysis. Failure to
     include or account for a confounding variable might cause a researcher to
     miss an underlying causality by considering only data showing no correlation.
5   The statement is not sensible. We cannot conclude that a correlation implies
     causality, regardless of the value of the correlation coefficient.
7   The statement is not sensible. Even though coincidence is ruled out, there
     might be a common underlying cause that could be a possible explanation.

Concepts and Applications

9   The causal connection is valid. As students spend more time and effort
     preparing for a test, the test grades tend to be higher.
11   This causal connection is valid. Alcohol is a depressant to the central
     nervous system, and it has several effects that include decreased reaction
     time. This is one important reason why drinking and driving is so dangerous.
13   (a)   Guideline 1
     (b)   Guidelines 2 and 5
     (c)   Guidelines 3 and 5

The headaches are associated with work days in some way. The headaches are not associated with Coke or possibly with caffeine. The headaches are possibly the result of bad ventilation in the building.

15   Smoking can only increase the risk already present.

17   This was an observational study. Later child bearing reflects an underlying cause. While it's possible that the conclusions are correct, there are other possible explanations for the findings. For example, it's also possible that the younger women lived during a time when having babies after age forty was less likely (by choice). It is still possible for them to live to be 100. Since this was an observational study, it is not possible to establish cause and effect.

19   Availability is not itself a cause. Social, economic, or personal conditions may also cause individuals to use the available weapons.

Chapter 7 Review Exercises

1

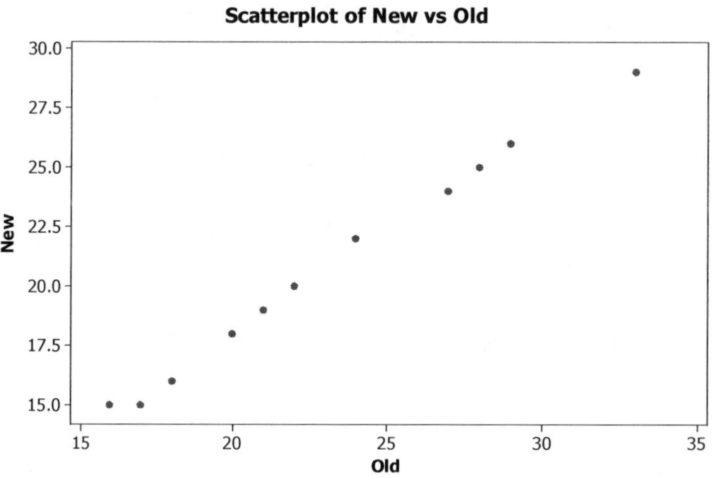

3   No. There is a strong linear relationship between the old and new ratings, but the old ratings did not cause the new ratings. Both were the result of the cars' performance.

5   The correlation could be the result of a coincidence; the correlation is due to a common cause; the correlation occurs because one variable has a liner effect on the other variable.

7   Correlation alone never implies causation (so we can't conclude that higher incomes cause more trips to the dentist), and, in this case, certainly more trips to the dentist do not cause higher incomes. Households with more disposable income can afford more trips to the dentist or can afford dental insurance that covers the costs of the trips.

9   The data values that were collected were uncorrelated. It's still possible that the variables represented by the data values are related in some non-linear way, i.e., the scatter plot forms a curve instead of a straight line.

Chapter 7 Quiz

1   Every possible correlation coefficient must lie between the values of –1 and 1.

3   Only statement a can be used to describe the relationship between the two variables. Statement d is not valid regardless of the strength of the correlation. Statement e is not valid since it is possible, but not ensured, that one of the variables is the direct cause of the other variable.

5 Yes. The data points all lie close to a straight line.

7 True.

9 False. There may be no pattern, or a pattern indicating some kind of relationship between two variables.

Example of the Computations of r and the Best-fit Line.

We provide here the details needed for the computation of r and the best-fit line for Exercise 19 of Section 7.1 and Exercise 13 of Section 7.3. Both exercises use the same data.

Speed Limit	Death Rate			
x	y	x^2	y^2	xy
55	3.0	3025	9.00	165.0
55	3.3	3025	10.89	181.5
55	3.4	3025	11.56	187.0
70	3.5	4900	12.25	245.0
55	4.1	3025	16.81	225.5
60	4.3	3600	18.49	258.0
55	4.7	3025	22.09	258.5
65	4.9	4225	24.01	318.5
60	5.1	3600	26.01	306.0
75	6.1	5625	37.21	457.5
605	**42.4**	**37075**	**188.32**	**2602.5**

For this data, the number of data points is n = 10. The totals for each column are shown in bold at the bottom of the column. Thus

$\sum x = 605$; $\sum y = 42.4$; $\sum x^2 = 37075$; $\sum y^2 = 188.32$; $\sum xy = 26025$
. The formula for r is given at the end of Section 7.1. Substituting the above values for the various sums, we have

$$r = \frac{n\sum xy - \sum x \sum y}{\sqrt{n\sum x^2 - (\sum x)^2}\sqrt{n\sum y^2 - (\sum y)^2}} = \frac{10 \times 2602.5 - 605 \times 42.4}{\sqrt{(10)(37075)} - \sqrt{(10)(188.32) - 42.4^2}} = 0.59.$$

$$r^2 = 0.59^2 = 0.35$$

$$s_x = \sqrt{\frac{n\sum x^2 - (\sum x)^2}{n-1}} = \sqrt{\frac{10\sum 370752 - (605)^2}{9}} = 7.246 \qquad s_y = \sqrt{\frac{n\sum y^2 - (\sum y)^2}{n-1}} = \sqrt{\frac{10\sum 188.32 - (42.4)^2}{9}} = 0.974$$

To find the best-fit line, we use the equations found at the end of Section 7.3. Substituting the various totals from the table above into those equations, we obtain:

$$m = r\frac{s_y}{s_x} = 0.59\frac{0.974}{7.246} = 0.07894. \qquad b = \bar{y} - m\bar{x} = 4.24 - 0.07894(60.5) = -0.536$$

Thus the equation of the best-fit line is

y = -0.536 + 0.07894x where y is the death rate and x is the speed limit.

To plot the line, we need to find two points on the line. Since any two points will suffice, we will choose $x = 55$ and $x = 75$. For $x = 55$, we have $y = -0.536 + 0.07894(55) = 3.81$, and for $x = 75$, we have $y = -0.536 + 0.07894(75) = 5.38$. The best-fit line can now be drawn by connecting the points (55, 3.81) and (75, 5.38).

Section 8.1
Statistical Literacy and Critical Thinking

1 The distribution of randomly selected digits from 0 to 9 is uniform. The distribution of sample means of 50 such digits is approximately normal.

3 The symbol $\bar{x}$ denotes the mean of a sample while μ denotes the mean of a population. Since the sample is a subset of the population, the sample mean and the population mean may be different.

5 The statement is not sensible. The larger size of the first sample does not guarantee that its sample mean is closer to the value of the population mean than that of the second sample. If the two samples are greatly different in size, it is likely that the mean of the larger sample is closer to the population mean, but it is still not guaranteed.

7 The statement is not sensible. The sample is a convenience sample and might not be representative of the county's population of cars for many reasons, including the location of the mall, the types of stores in the mall, the incomes of the people who shop there, and the time of day at which the sample is taken.

Concepts and Applications

9 The best estimate of the population proportion for the Northeast is the same as the sample proportion: $\hat{p}$ = 0.19. The given sample proportion is not likely to be a good estimate of the proportion for the population of all children in the United States simply because the sample was not taken from the entire U.S. There are regional factors that could have a strong influence on that proportion in different parts of the country.

11 a) The sample mean is (12.19 – 12.00)/0.02 = 9.5 standard deviations away from the mean of the distribution of sample means.

  b) Based on Table 5.1, the probability is very small, less than 0.0001 or 0.01%.

  c) No. It appears that the cans are being filled with an amount that is greater than 12.00 ounces, so consumers are getting more than the amount stated on the label.

13 a) The population proportion is p = 6523/12345 = 0.528.

  b) The sample proportion is $\hat{p}$ = 245/500 = 0.49.

  c) The sample proportion is a little too low.

15 The sample proportion is $\hat{p}$ = 73/150 = 0.4867. This is the best estimate of the population proportion, so we estimate that 0.4867 x 1608 = 783 people traveled from abroad. We would be more confident of the estimate if we sampled 300 people instead of 150 since the estimate $\hat{p}$ becomes more reliable as the sample size increases.

17 a) $z = \dfrac{\text{sample proportion} - \text{population proportion}}{\text{standard deviation}} = \dfrac{0.32-0.34}{0.03} = -0.67$, so the sample proportion is 0.67 standard deviations below the mean of the sampling distribution.

  b) From Table 5.1 (interpolating), z = -0.67 corresponds to the 25.14 percentile, so the probability of a sample proportion less than 0.32 is 0.2514.

19   a)

Sample	Sample Mean
1,1	1.0
1,2	1.5
1,5	3.0
2,1	1.5
2,2	2.0
2,5	3.5
5,1	3.0
5,2	3.5
5,5	5.0

b)   The mean of the sample means in part (a) is
$(1.0 + 1.5 + \ldots + 5.0)/9 = 24/9 = 8/3 = 2.7$.

c)   The mean of the population is $(1 + 2 + 5)/3 = 8/3 = 2.7$, which is the same as the mean of the sample means in part (a). Yes, these means will always be equal.

21   (a)   On the right side of the following table, we show the sample means for each of the samples of size n = 2.

Sample		Mean
5	5	5
52	5	28.5
60	5	32.5
5	52	28.5
52	52	52
60	52	56
5	60	32.5
52	60	56
60	60	60

(b)   The mean of the sample means is $(5 + 28.5 + \ldots + 60)/9 = 39$.

(c)   The mean of the population is $(5 + 52 + 60)/3 = 39$, so the mean of the sample means is the same as the mean of the population. This is always the case.

Section 8.2
Statistical Literacy and Critical Thinking

1   We have 95% confidence that the limits of 8.0518 g and 8.0902 g actually do contain the true population mean weight of dollar coins. We expect that 95% of such samples will result in confidence interval limits that do contain the population mean.

3   The media often omit reference to the confidence level, which is typically 95%. The word "mean" should be used instead of the word "average" since there are other kinds of averages besides the mean.

5   The statement is not sensible. 95% of all sample means will fall within 2 standard deviations of the population mean, not 2 standard deviation os one sample mean.

7   The statement is generally true. As the sample size increases, the margin of error tends to decrease (as long as the standard deviation does not increase).

Concepts and Applications

9   The margin of error is $E \cong \frac{2s}{\sqrt{n}} = \frac{2 \times 2.2}{\sqrt{49}} = 0.6$ cm.   The approximate 95% confidence interval is $\bar{x} \pm E = 25.2 \pm 0.6 = 24.6 \, to \, 25.8$ cm.

11 The margin of error is $E \cong \frac{2s}{\sqrt{n}} = \frac{2 \times 2.0}{\sqrt{100}} = 0.4$ ft.   The approximate 95% confidence interval is $\bar{x} \pm E = 8.0 \pm 0.4 = 7.6 \, to \, 8.4$ ft.

13 $n = \left[\frac{2\sigma}{E}\right]^2 = \left[\frac{2 \times 20}{5}\right]^2 = 64.$

15 $n = \left[\frac{2\sigma}{E}\right]^2 = \left[\frac{2 \times 155.2}{3.5}\right]^2 = 7865.2$, so we use n = 7866.

17 $n = \left[\frac{2\sigma}{E}\right]^2 = \left[\frac{2 \times 1.7}{0.25}\right]^2 = 184.96$, so we use n = 185.

19 $n = \left[\frac{2\sigma}{E}\right]^2 = \left[\frac{2 \times 15}{3}\right]^2 = 100.$

21 The population mean μ is estimated by the sample mean $\overline{x}$ = 5.639 g.   The margin of error is $E \cong \frac{2s}{\sqrt{n}} = \frac{2 \times 0.062}{\sqrt{40}} = 0.020 \, g$.   The approximate 95% confidence interval is $\bar{x} \pm E = 5.639 \pm 0.020 = 5.619 \, g \, to \, 5.659 \, g$.

23 The population mean μ is estimated by the sample mean $\overline{x}$ = 5.15 years.   The margin of error is $E \cong \frac{2s}{\sqrt{n}} = \frac{2 \times 1.68}{\sqrt{4400}} = 0.05 \, years$.   The approximate 95% confidence interval is $\bar{x} \pm E = 5.15 \pm 0.05 = 5.10 \, to \, 5.20$ years.

25 Use the software of your choice to find the mean and standard deviation of the sample of n = 36 weights.   You should find $\overline{x}$ = 183.58 and s = 112.95.   The margin of error is $E \cong \frac{2s}{\sqrt{n}} = \frac{2 \times 112.95}{\sqrt{36}} = 37.65 \, lbs$ and the 95% confidence interval is $\bar{x} \pm E = 183.58 \pm 37.65 = 146 \, to \, 221$ pounds or 146 < μ < 221 lb.

27 Use the software of your choice to find the mean and standard deviation of the sample of n = 31 family sizes.
   a) $\overline{x}$ = 3.10
   b)   s = 1.423
   c)   The best estimate for the mean family size for the population of all American families is the sample mean, 3.10.
   d)   The margin of error is $E \cong \frac{2s}{\sqrt{n}} = \frac{2 \times 1.423}{\sqrt{31}} = 0.51$ and the 95% confidence interval is $\bar{x} \pm E = 3.10 \pm 0.51 = 2.59 \, to \, 3.61$ or 2.59 < μ < 3.61.
   e)   We can be 95% confident that the interval from 2.59 to 3.61 contains the population mean μ.   That is, we expect 95% of confidence intervals computed from such samples to contain the population mean.   Since the sample is taken entirely from one neighborhood, this should be a reliable estimate for homes in the neighborhood, but it may not be very reliable if the population of interest is larger than the neighborhood.

This problem can also be done by hand although it is somewhat tedious.   We create a table of x values, x – $\overline{x}$ values and the squares of those differences as shown following.

x	$x - \overline{x}$	$(x - \overline{x})^2$
2	−1.1	1.21
3	−0.1	0.01
6	2.9	8.41
5	1.9	3.61
4	0.9	0.81
2	−1.1	1.21
3	−0.1	0.01
3	−0.1	0.01
1	−2.1	4.41
2	−1.1	1.21
3	−0.1	0.01
2	−1.1	1.21
3	−0.1	0.01
4	0.9	0.81
5	1.9	3.61
3	−0.1	0.01
1	−2.1	4.41
3	−0.1	0.01
3	−0.1	0.01
4	0.9	0.81
7	3.9	15.21
3	−0.1	0.01
2	−1.1	1.21
3	−0.1	0.01
2	−1.1	1.21
2	−1.1	1.21
3	−0.1	0.01
4	0.9	0.81
1	−2.1	4.41
5	1.9	3.61
2	−1.1	1.21
96		**60.71**

Dividing the total in the first column by 31, we find that $\overline{x}$ = 3.1.
Dividing the total in the third column by n − 1 = 30, and then taking the
square root of the result, we find that s = 1.423.

Section 8.3
Statistical Literacy and Critical Thinking

1   We have 95% confidence that the limits of 0.393 and 0.553 actually do contain
the true population proportion. We expect that 95% of such samples will result
in confidence interval limits that do contain the population proportion.

3   The media often omit reference to the confidence level, which is typically 95%.

5   The statement is not sensible. The correct interpretation is that we have 95%
confidence because we know that 95% of all such intervals will contain the true
population proportion p.

7   The statement is sensible. We can see from the expression for E that larger
values of n result in smaller values of E. Also, common sense suggests that
larger samples are likely to result in better and more accurate (that is,
smaller errors) estimates of a population proportion if $\hat{p}$ doesn't change.

Concepts and Applications

9 The approximate margin of error is $E = 2\sqrt{\dfrac{\hat{p}(1-\hat{p})}{n}} = 2\sqrt{\dfrac{0.4(1-0.4)}{1000}} = 0.031$.   The 95% confidence interval is $\hat{p} \pm E = 0.4 \pm 0.031 = 0.369 \text{ to } 0.431$. $0.369 < p < 0.431$.

11 The approximate margin of error is $E = 2\sqrt{\dfrac{\hat{p}(1-\hat{p})}{n}} = 2\sqrt{\dfrac{0.1(1-0.1)}{550}} = 0.0256$.   The 95% confidence interval is $\hat{p} \pm E = 0.1 \pm 0.0257 = 0.0743 \text{ to } 0.1257$; $0.074 < p < 0.126$.

13 $n = \dfrac{1}{E^2} = \dfrac{1}{0.04^2} = 625$

15 $n = \dfrac{1}{E^2} = \dfrac{1}{0.015^2} = 4444.4$, so we use n = 4445.

17 The approximate margin of error is $E = 2\sqrt{\dfrac{\hat{p}(1-\hat{p})}{n}} = 2\sqrt{\dfrac{0.15(1-0.15)}{5000}} = 0.010$.   The 95% confidence interval is $\hat{p} \pm E = 0.15 \pm 0.010 = 0.140 \text{ to } 0.160$; or $0.140 < p < 0.160$.

19 The approximate margin of error is $E = 2\sqrt{\dfrac{\hat{p}(1-\hat{p})}{n}} = 2\sqrt{\dfrac{0.8(1-0.8)}{1400}} = 0.021$.   The 95% confidence interval is $\hat{p} \pm E = 0.800 \pm 0.021 = 0.779 \text{ to } 0.821$; or $0.779 < p < 0.821$.

21 $\hat{p} = \dfrac{127}{152} = 0.836$.   The margin of error is $E = 2\sqrt{\dfrac{\hat{p}(1-\hat{p})}{n}} = 2\sqrt{\dfrac{0.836(1-0.836)}{152}} = 0.060$.   The 95% confidence interval is $\hat{p} \pm E = 0.836 \pm 0.060 = 0.775 \text{ to } 0.896$; or $0.775 < p < 0.896$.

23 a)   The sample proportion is 0.98.

b)   From this, the margin of error is $E = 2\sqrt{\dfrac{\hat{p}(1-\hat{p})}{n}} = 2\sqrt{\dfrac{0.98(1-0.98)}{400}} = 0.014$.   The 95% confidence interval is $\hat{p} \pm E = 0.98 \pm 0.014 = 0.966 \text{ to } 0.994$; or $0.966 < p < 0.994$.

c)   No.   To be a random sample, all films must have an equal chance of being chosen.

25 For the first poll, $\hat{p} = \dfrac{780}{1500} = 0.520$.   The margin of error is $E = 2\sqrt{\dfrac{\hat{p}(1-\hat{p})}{n}} = 2\sqrt{\dfrac{0.520(1-0.520)}{1500}} = 0.026$.   The 95% confidence interval is $\hat{p} \pm E = 0.520 \pm 0.026 = 0.494 \text{ to } 0.546$; or $0.494 < p < 0.546$.

For the second poll, $\hat{p} = \dfrac{1285}{2500} = 0.514$.   The margin of error is $E = 2\sqrt{\dfrac{\hat{p}(1-\hat{p})}{n}} = 2\sqrt{\dfrac{0.514(1-0.514)}{2500}} = 0.020$.   The 95% confidence interval is $\hat{p} \pm E = 0.514 \pm 0.020 = 0.494 \text{ to } 0.534$; or $0.494 < p < 0.534$.

For the third poll, $\hat{p} = \dfrac{1802}{3500} = 0.515$. The margin of error is $E = 2\sqrt{\dfrac{\hat{p}(1-\hat{p})}{n}} = 2\sqrt{\dfrac{0.515(1-0.515)}{3500}} = 0.017$.   The 95% confidence interval is $\hat{p} \pm E = 0.515 \pm 0.017 = 0.498 \text{ to } 0.532$; or $0.498 < p < 0.532$.

Because all of these confidence intervals include values below 0.5, Martinez cannot be confident of winning a majority.

27 a)   The 95% confidence interval for the true population proportion approving the President's performance is $0.54 \pm 0.04 = 0.50 \text{ to } 0.58$.

b)   Since $E = 2\sqrt{\dfrac{\hat{p}(1-\hat{p})}{n}} = 2\sqrt{\dfrac{0.54(1-0.54)}{n}} = 0.04$, we conclude that $\sqrt{\dfrac{0.54(1-0.54)}{n}} = 0.02$ or $\dfrac{(0.54)(0.46)}{n} = 0.0004$.   This implies that $n = \dfrac{(0.54)(0.46)}{0.0004} = 621$.   The sample size may have been determined as $1/E^2 = 1/0.04^2 = 625$.   The difference is small.

Chapter 8 Review Exercises

1 a) $\hat{p} = \frac{589}{745} = 0.791$.

    b)   The margin of error is $E = 2\sqrt{\frac{\hat{p}(1-\hat{p})}{n}} = \sqrt{\frac{0.791(1-0.791)}{745}} = 0.030$.   The 95% confidence interval is $\hat{p} \pm E = 0.791 \pm 0.030 = 0.761 \; to \; 0.820$; or $0.761 < p < 0.820$.

    c)   E = 0.030

    d)   We have 95% confidence that the limits of 0.761 and 0.820 contain the true value of the population proportion. If such samples of size 745 were randomly selected many times, the resulting confidence interval limits would contain the true population proportion in 95% of those samples.

    e)   The distribution of the sample proportions should be approximately normal or bell-shaped.

    f)   n = $1/E^2$ = $1/0.02^2$ = 2,500.

3 a)   The margin of error is $E = \frac{2s}{\sqrt{n}} = \frac{2 \times 65.2}{\sqrt{40}} = 20.6$.   The 95% confidence interval is $\bar{x} \pm E = 279.5 \pm 20.6 = 258.9 \; to \; 300.1$.

    b)   The margin of error, from part (a), is E = 20.6.

    c)   We can be 95% confident that the interval from 258.9 to 300.1 contains the population mean white blood cell count. If such samples of size 40 were randomly selected many times, the resulting confidence interval limits would contain the true population mean in 95% of those samples.

    d) $n = \left[\frac{2s}{E}\right]^2 = \left[\frac{2 \times 65.2}{10}\right]^2 = 171$

Chapter 8 Quiz

1   The distribution of the sample means is approximately a normal distribution.

3   The symbol $\hat{p}$ represents the sample proportion.

5   The best estimate of the population mean is the mean of the sample – (b).

7   The margin of error is one-half of the length of the confidence interval. Since the length of the interval is 98.6 – 98.0 = 0.6, the margin of error is 0.3.

9   The 95% confidence interval is $\hat{p} \pm E = 0.720 \pm 0.025 = 0.695 \; to \; 0.745$.

<u>**Section 9.1**</u>
Statistical Literacy and Critical Thinking

1 A hypothesis test is a standard procedure for testing a claim about the value of a population parameter.

3 The alternative hypothesis could be $\mu < 195$, $\mu > 195$, or $\mu \neq 195$.

5 The statement is not sensible. A hypothesis test cannot be used to prove that a parameter has a particular value. It can be used to say that a particular value is unlikely or that the particular value is reasonable, but never that a particular value is absolutely correct.

7 The statement is sensible. A null hypothesis always states a specific value of a parameter by using an equal sign. The expression $\mu < 130$ mg/dL is in the form of a correct alternative hypothesis.

9 This statement is sensible. The null hypothesis should state that $p = 0.06$.

11 This statement is not sensible. A P-value of 0.45 indicates that an event has occurred that could happen almost half of the time. This means that a very common event has occurred. In that case, the experiment does not provide support for the alternative hypothesis.

Concepts and Applications

13 a) Getting 22 girls in 40 children is close to what we would expect by chance, so it is not significant.
   b) Getting 35 girls in 40 children is significant since this is many more than we would expect by chance.
   c) Part (b). This very small P-value would be associated with a rare event, 35 girls in 40 babies.

15 H_0: $\mu = 250$ calories; H_a: $\mu \neq 250$ calories. The two possible conclusions are that there is sufficient sample evidence to reject the claim that the mean caloric content is equal to 250 calories, or that there is not sufficient sample evidence to reject the claim that the mean caloric content is 250 calories.

17 H_0: $p = 0.7$; H_a: $p > 0.7$. The two possible conclusions are that there is not sufficient sample evidence to support the claim that the proportion of girls is greater than 0.7, or that there is sufficient sample evidence to support the claim that the proportion of girls is greater than 0.7.

19 H_0: $\mu = 10$ oz; H_a: $\mu \neq 10$ oz. The two possible conclusions are that there is not sufficient sample evidence to reject the claim that the mean amount supplied is equal to 10 oz, or that there is sufficient sample evidence to reject the claim that the mean amount supplied is equal to 10 oz.

21 H_0: $p = 0.5$; H_a: $p > 0.5$. The two possible conclusions are that there is not sufficient sample evidence to support the claim that the proportion of students is greater than 0.5, or that there is sufficient sample evidence to support the claim that the proportion of students is greater than 0.5.

In Exercises 23, 25, and 27, we are testing for a bias against male babies. The null hypothesis reflects the situation if there is no bias and the alternative hypothesis reflects the situation if a bias exists.

23 No. The P-value of 0.382 is greater than the significance level of 0.05, and this high value suggests that there is a high likelihood of getting 48 or fewer males by chance.

25 No. The P-value of 0.028 is greater than the significance level of 0.01. While the occurrence of 40 or fewer male babies by chance is not very high, it is still larger than the significance level.

27 Yes. The P-value of 0.002 is less than the significance level of 0.01. This low value suggests that there is only a small possibility of getting 35 or fewer males by chance.

Section 9.2
Statistical Literacy and Critical Thinking

1 n represents the sample size; $\bar{x}$ represents the mean of the sample; s represents the standard deviation of the sample; σ represents the standard deviation of the population; and μ represents the mean of the population.

3 A Type I error is made when the null hypothesis is rejected, but it is actually true. A Type II error is made when failing to reject a null hypothesis that is actually false.

5 The statement is not sensible. The significance level is chosen prior to collecting data and is frequently a number like 0.05 or 0.01. The P-value is found after the sample results have been obtained and is dependent on the sample size, the sample statistic, and the hypotheses about the population parameter.

7 This statement is sensible. Small P-values (near 0) lead to support of the alternative hypothesis. A P-value of 0.001 is very small and would lead to rejection of the null hypothesis or support of the alternative hypothesis.

9 This statement makes sense. The significance level is the probability of making a Type I error.

11 This statement does not make sense. If the significance level were zero, the null hypothesis could never be rejected, even when it is ridiculous. In that case, there would be no point to conducting the experiment in the first place; the hypothesis test would be meaningless.

Concepts and Applications

13 $z = \frac{\bar{x} - \mu}{\sigma/\sqrt{n}} = \frac{70 - 75}{15/\sqrt{100}} = -3.33$. Since z < -1.645, we reject the null hypothesis.
There is sufficient sample evidence to support the alternative hypothesis that μ < 75.

15 $z = \frac{\bar{x} - \mu}{\sigma/\sqrt{n}} = \frac{14 - 12}{2/\sqrt{64}} = 8.0$. Since z > 1.645, we reject the null hypothesis. There is sufficient sample evidence to support the alternative hypothesis that μ > 12.

17 $z = \frac{\bar{x} - \mu}{\sigma/\sqrt{n}} = \frac{2.58 - 2.55}{0.29/\sqrt{100}} = 1.03$. Since -1.96 < z < 1.96, we do not reject the null hypothesis. There is not sufficient sample evidence to support the alternative hypothesis that $\mu \neq 2.55$.

19 $z = \frac{\bar{x} - \mu}{\sigma/\sqrt{n}} = \frac{0.75 - 0.88}{0.18/\sqrt{50}} = -5.11$. Since z < -1.96, we reject the null hypothesis.
There is sufficient sample evidence to support the alternative hypothesis that $\mu \neq 0.88$.

In Exercises 21-33, most of the z values are given to tenths and so it is easy to find the P-values using Table 5.1. In practice, many z values are between those showing in Table 5.1. In Excel, there is a function that can be used to find the P-value. Assuming that the z value is in cell A1, enter the following expressions in a different cell to find the P-value.

Alternative	Left-tail	Right-tail	Both Tails
Expression	`=NORMSDIST(A1)`	`=1-NORMSDIST(A1)`	`=2*(NORMSDIST(ABS(A1)))`

21 Using Table 5.1, P-value = P(z < -0.4) = 0.3446. Since P-value > 0.05, there is not sufficient sample evidence to support the alternative hypothesis.

23 Using Table 5.1, P-value = P(z > 1.9) = 1 - P(z < 1.9) = 1 - 0.9713 = 0.0287. Since P-value < 0.05, there is sufficient sample evidence to support the alternative hypothesis.

25 Using Table 5.1, P-value = 2P(z < -1.6) = 2(0.0548) = 0.1096. Since P-value > 0.05, there is not sufficient sample evidence to support the alternative hypothesis.

27 Using Table 5.1, P-value = 2P(z > 1.7) = 2(1 - 0.9554) = 0.0892. Since P-value > 0.05, there is not sufficient sample evidence to support the alternative hypothesis.

29 Using Table 5.1, P-value = P(z > 2.7) = 1 - P(z < 2.7) = =1 - 0.9965 = 0.0035. Since P-value < 0.05, there is sufficient sample evidence to support the alternative hypothesis.

31 Using Table 5.1, P-value = P(z < -2.1) = 0.0179. Since P-value < 0.05, there is sufficient sample evidence to support the alternative hypothesis.

33 Using Table 5.1, P-value = 2P(z > 0.15) = 2(1 - 0.5596) = 0.8808. Since P-value > 0.05, there is not sufficient sample evidence to support the alternative hypothesis.

35 In order to reject the null hypothesis, the z value will need to be positive. Since z = -1.0, there is no need to find the P-value exactly. It will be greater than 0.5.

37 H_0: $\mu = 7.5$; H_a: $\mu < 7.5$. Since the sample size is large, we'll assume that $\sigma = s$. Then $z = \frac{\bar{x} - \mu}{\sigma/\sqrt{n}} = \frac{7.01 - 7.50}{3.74/\sqrt{100}} = \frac{-0.49}{0.374} = -1.31$. From Table 5.1, the P-value = P(z < -1.31) = 0.0968. Since P-value > 0.05, there is not sufficient evidence to support the alternative hypothesis that the mean time of ownership for all cars is less than 7.5 years.

39 H_0: $\mu = 12.0$; H_a: $\mu \neq 12.0$. Since the sample size is large, we'll assume that $\sigma = s$. Then $z = \frac{\bar{x} - \mu}{\sigma/\sqrt{n}} = \frac{12.19 - 12.00}{0.11/\sqrt{36}} = \frac{0.19}{0.0183} = 10.36$. From Table 5.1, the P-value = 2P(z > 10.36) = 2(1 - 1.0000) = 2(0.0000) = 0.0000. Since P-value < 0.05, there is sufficient sample evidence to support the alternative hypothesis that the mean package weight is different from 12.0 oz.

41 H_0: $\mu = 600$ mg; H_a: $\mu \neq 600$ mg. Since the sample size is large, we'll assume that $\sigma = s$. Then $z = \frac{\bar{x} - \mu}{\sigma/\sqrt{n}} = \frac{589 - 600}{21/\sqrt{65}} = \frac{-11}{2.605} = -4.2$. From Table 5.1, the P-value = 2P(z < -4.2) = 2(0.0002) = 0.0004. Since P-value < 0.05, there is sufficient sample evidence to support the alternative hypothesis that the mean amount of acetominophen is different from 600 mg.

43 H_0: $\mu = 21.4$; H_a: $\mu < 21.4$. Since the sample size is large, we'll assume that $\sigma = s$. Then $z = \frac{\bar{x} - \mu}{\sigma/\sqrt{n}} = \frac{19.8 - 21.40}{3.5/\sqrt{40}} = \frac{-1.6}{.553} = -2.89$. From Table 5.1, the P-value = P(z < -2.89) = 0.0019. Since P-value < 0.05, there is sufficient evidence to support the alternative hypothesis that the mean mileage per gallon of all SUVs is less than 21.4 miles per gallon.

45 H_0: $\mu = 0.21$; H_a: $\mu > 0.21$. Since the sample size is large, we'll assume that $\sigma = s$. Then $z = \frac{\bar{x} - \mu}{\sigma/\sqrt{n}} = \frac{0.83 - 0.21}{.24/\sqrt{32}} = \frac{0.62}{.0424} = 14.6$. From Table 5.1, the P-value = P(z > 14.6) < 1 - 0.9998 = 0.0002. Since P-value < 0.05, there is sufficient sample evidence to support the alternative hypothesis that the mean score of self-identified buyers is greater than 0.21.

47 H_0: $\mu = 3.39$ kg; H_a: $\mu \neq 3.39$ kg. Since the sample size is large, we'll assume that $\sigma = s$. Then $z = \frac{\bar{x} - \mu}{\sigma/\sqrt{n}} = \frac{3.67 - 3.39}{0.66/\sqrt{121}} = \frac{0.28}{.06} = 4.67$. From Table 5.1, the P-value = 2P(z > 4.47) < 2(1 - 0.9998) = 0.0004. Since P-value < 0.05, there is sufficient sample evidence to support the alternative hypothesis that the mean birth weight of male babies born with the vitamin supplement is not equal to 3.39 kg.

49 H_0: μ = 24 months; H_a: μ < 24 months. Since the sample size is large, we'll assume that σ = s. Then $z = \frac{\bar{x} - \mu}{\sigma/\sqrt{n}} = \frac{22.1 - 24.0}{8.6/\sqrt{70}} = \frac{-1.9}{1.028} = -1.85$. From Table 5.1, the P-value = P(z < -1.85) = 0.0323. Since P-value < 0.05, there is sufficient evidence to support the alternative hypothesis that the mean prison term for convicted embezzlers is less than 24 months.

51 Type I error: Reject the claim that the patient is free of the disease when the patient is actually free of the disease; i.e., the patient is diagnosed with a disease he does not have.
Type II error: Fail to reject the claim that the patient is free of the disease when the patient is actually not free of the disease; i.e., the patient is not diagnosed with a disease he has.

53 Type I error: Reject the claim that the lottery is fair when the lottery is actually fair.
Type II error: Fail to reject the claim that the lottery is fair when the lottery is actually biased.

Section 9.3
Statistical Literacy and Critical Thinking

1 The symbol p represents the proportion in a population; $\hat{p}$ represents the proportion in a sample; and a P-value is the probability of getting a sample result that is at least as extreme as the sample result actually obtained.

3 No. The sample is a voluntary response sample, self-selected, and is not a simple random sample. It is likely that those who responded are not representative of the general population. They certainly can not be said to represent the many people who do not use the internet at all.

5 This statement makes sense. The null hypothesis must specify a single value for p. The alternative hypothesis corresponding to a majority is that p > 0.5.

7 This statement is not correct. For a two-tailed test, the P-value is twice the area to the right of z if z is positive or twice the area to the left of z if z is negative.

Concepts and Applications

9 H_0: p = 0.5; H_a: p > 0.5 From the sample, $\hat{p}$ = 205/400 = 0.5125. Then $z = \frac{\hat{p} - p}{\sqrt{\frac{p(1-p)}{N}}} = \frac{0.5125 - 0.5}{\sqrt{\frac{0.5(1-0.5)}{400}}} = \frac{0.0125}{0.025} = 0.5$. From Table 5.1, the P-value = P(z > 0.5) = 1 - 0.6915 = 0.3085. Since P-value > 0.05, there is not sufficient evidence to support the alternative hypothesis that a majority of the voters support the candidate.

11 H_0: p = 0.5; H_a: p < 0.5 From the sample, $\hat{p}$ = 0.44. Then $z = \frac{\hat{p} - p}{\sqrt{\frac{p(1-p)}{n}}} = \frac{0.44 - 0.50}{\sqrt{\frac{0.5(1-0.5)}{1015}}} = \frac{-0.06}{0.0157} = -3.82$. From Table 5.1, the P-value = P(z < -3.82) < 0.0002.
Since P-value < 0.05, there is sufficient evidence to support the alternative hypothesis that fewer than half of all teenagers in the population feel that grades are the greatest source of pressure.

13 H_0: p = 0.099; H_a: p < 0.099 From the sample, $\hat{p}$ = 98/1050 = 0.093. Then $z = \frac{\hat{p} - p}{\sqrt{\frac{p(1-p)}{n}}} = \frac{0.093 - 0.099}{\sqrt{\frac{0.099(1-0.099)}{1050}}} = \frac{-0.006}{0.0092} = -0.61$. From Table 5.1, the P-value = P(z < -0.61) = 0.2743. Since P-value > 0.05, there is not sufficient evidence to support the alternative hypothesis that illegal drug use in her school is less than the current national average.

15 H_0: p = 0.5; H_a: p > 0.5 From the sample, $\hat{p}$ = 0.51. Then $z = \dfrac{\hat{p}-p}{\sqrt{\frac{p(1-p)}{n}}} =$

$\dfrac{0.51-0.50}{\sqrt{\frac{0.50(1-0.50)}{276000}}} = \dfrac{-0.01}{0.000952} = 10.51$. From Table 5.1, the P-value = P(z > 10.51) < 1 – 0.9998 = 0.0002. Since P-value < 0.05, there is sufficient sample evidence to support the alternative hypothesis that over half of all first-year students believe that abortion should be legal.

17 H_0: p = 0.53; H_a: p ≠ 0.53 From the sample, $\hat{p}$ = 0.54. Then $z = \dfrac{\hat{p}-p}{\sqrt{\frac{p(1-p)}{n}}} =$

$\dfrac{0.54-0.53}{\sqrt{\frac{0.53(1-0.53)}{3600}}} = \dfrac{0.01}{0.0083} = 1.20$. From Table 5.1, the P-value = 2P(z > 1.20) = 2(1 – 0.8849) = 2(0.1151) = 0.2302. Since P-value > 0.05, there is not sufficient sample evidence to support the alternative hypothesis that the percentage of households in the U.S. using natural gas has changed.

Chapter 9 Review Exercises

1   a) H_0: μ = 12.0 oz

  b) H_a: μ > 12.0 oz

  c)   The standard deviation of the sampling distribution of the mean is $\dfrac{\sigma}{\sqrt{n}} \approx \dfrac{s}{\sqrt{n}} = \dfrac{0.11}{\sqrt{36}} = 0.0183$; the standard score for the sample means is $z = \dfrac{\bar{x}-\mu}{\sigma/\sqrt{n}} = \dfrac{12.19-12.00}{0.0183} = 10.4$

  d)   Since the test is right-tailed, the critical value is 1.645.

  e)   The P-value is the probability that z is greater than 10.4, which is less than 0.0002.

  f)   Since the P-value is less than 0.05, we reject the null hypothesis and claim that cans of Coke have a mean amount of cola greater than 12 ounces.

  g)   A type I error would result if we concluded that the population mean was greater than 12 ounces when, in fact, it was not.

  h)   A type II error would result if we did not conclude that the population mean was greater than 12 ounces when, in fact, it is greater than 12 ounces.

  i)   The P-value would be the probability that z is greater than 10.4 or less than -10.4. Since the probability that z is greater than 10.4 is less than 0.0002, the P-value is less than 2 x 0.0002 = 0.0004.

3   (a)   If p is the proportion of those getting a job through networking, then the null hypothesis is that p = 0.5. The alternative hypothesis is that p must be greater than 0.5. The critical value is 1.645. For these data, $\hat{p}$ = 429/703 = 0.61. Then $z = \dfrac{\hat{p}-p}{\sqrt{\frac{p(1-p)}{N}}} = \dfrac{0.61-0.50}{\sqrt{\frac{0.5(1-0.5)}{703}}} = \dfrac{.11}{0.0189} = 5.85$.

  There is sufficient evidence to reject the null hypothesis and claim that the majority get jobs through networking.

  (b)   With $\hat{p}$ =0.61, there is no way that the data could support a claim that p < 0.5. We would need a value of $\hat{p}$ that is less than 0.5.

Chapter 9 Quiz

1   μ > 1126 cm^3

3   p ≠ 0.6

5   μ = 110

7   Since the P-value is less than 0.05, we conclude that the alternative hypothesis is supported by the sample data. We reject the null hypothesis. The data support the claim that the mean is greater than 110.

9   The data either support the claim that the IQ of professional comedians is greater than 110, or they do not support the claim.

Section 10.1
Statistical Literacy and Critical Thinking

1 No. The sample is a convenience sample, not a simple random sample. It is not likely to be representative of the population.

3 There is no difference. The sampling distribution of the t statistic was discovered by W.S. Gosset and published in 1908 under the pen name of "Student." Since then, it has been known as the Student t.

5 This statement does not make sense. Although the t distribution can be used with small samples, those samples must come from a normal or nearly normal population. If there is any indication from the sample that the data are not from a normal population, the t distribution should not be used for making inferences about the mean of the population. The larger the sample gets, the more non-normality that can be tolerated.

7 This statement makes sense. The use of the t distribution for small samples requires that the population have a normal distribution.

Concepts and applications

9 The degrees of freedom is n − 1 = 16 − 1 = 15. From Table 10.1, the t value is 2.131. The margin of error is $E = t\frac{s}{\sqrt{n}} = 2.131\frac{10}{\sqrt{16}} = 5.3$. The 95% confidence interval is $\bar{x} \pm E = 130 \pm 5.3 = 124.7 \text{ to } 135.3$.

11 The degrees of freedom is n − 1 = 35 − 1 = 34. From Table 10.1, the t value is 2.032. The margin of error is $E = t\frac{s}{\sqrt{n}} = 2.032\frac{0.7}{\sqrt{35}} = 0.240$. The 95% confidence interval is $\bar{x} \pm E = 14.5 \pm 0.2 = 14.3 \text{ to } 14.7$.

13 a) The degrees of freedom is n − 1 = 20 − 1 = 19. From Table 10.1, the t value is 2.093. The margin of error is $E = t\frac{s}{\sqrt{n}} = 2.093\frac{5629}{\sqrt{20}} = 2634$.

The 95% confidence interval is $\bar{x} \pm E = 9004 \pm 2634 = \$6370 \text{ to } \$11{,}638$.

b) Use the upper confidence limit of $11,638 as the worst possible hospital cost to the insurance company. One could be even more conservative by using a 99% confidence interval (which would be wider). The upper confidence limit would be higher yet.

15 a) We must first find the sample mean and standard deviation. The table below will be helpful.

Nitrogen oxide x	$x - \bar{x}$	$(x - \bar{x})^2$
0.06	−0.061	0.0036
0.11	−0.011	0.0001
0.16	0.039	0.0016
0.15	0.029	0.0009
0.14	0.019	0.0004
0.08	−0.041	0.0016
0.15	0.029	0.0009
0.121=Mean		0.0091

The sample mean is 0.121. The standard deviation is $s = \sqrt{0.0091/6} = 0.0389$.

The degrees of freedom is n − 1 = 7 − 1 = 6. From Table 10.1, the t value is 2.447. The margin of error is $E = t\frac{s}{\sqrt{n}} = 2.447\frac{0.0389}{\sqrt{6}} = 0.0360$. The 95% confidence interval is $\bar{x} \pm E = 0.121 \pm 0.036 = 0.085 \text{ to } 0.157$.

b)   Since the confidence interval does not contain 0.165, we can conclude that the mean is not equal to 0.165 g/mile.

17 H_0: $\mu = 0.3$; H_a: $\mu < 0.3$. The degrees of freedom is $n - 1 = 16 - 1 = 15$. The critical value from Table 10.1 is -1.753. $t = \dfrac{\bar{x}-\mu}{s/\sqrt{n}} = \dfrac{0.295-0.3}{0.168/\sqrt{16}} = -0.119$. Since this value is not less than -1.753, do not reject the null hypothesis. The data do not support the claim that the mean amount of sugar in all cereals is less than 0.3 g. The P-value can easily be obtained using software. For example, in Excel, in any cell enter the expression =TDIST(0.119,15,1) and press ENTER. The result is 0.4534. The numbers in the parentheses are the t statistic (entered as a positive number), the degrees of freedom, and a 1 to indicate a one-sided test.

19 H_0: $\mu = 2.5$; H_a: $\mu \neq 2.5$. The degrees of freedom is $n - 1 = 30 - 1 = 29$. The critical values from Table 10.1 are -2.045 and 2.045. $t = \dfrac{x-\mu}{s/\sqrt{n}} = \dfrac{2.4991-2.5}{0.01648/\sqrt{30}} = -0.299$. Since this value is not less than -2.045, do not reject the null hypothesis. The data do not support the claim that the mean penny weight is different from 2.5 g. The P-value can easily be obtained using software. For example, in Excel, in any cell enter the expression =TDIST(0.299,29,2) and press ENTER. The result is 0.7671. The numbers in the parentheses are the t statistic (entered as a positive number), the degrees of freedom, and a 2 to indicate a two-sided test.

21 H_0: $\mu = 3.39$; H_a: $\mu \neq 3.39$. The degrees of freedom is $n - 1 = 16 - 1 = 15$. The critical values from Table 10.1 are -2.131 and 2.131. $t = \dfrac{\bar{X}-\mu}{s/\sqrt{n}} = \dfrac{3.675-3.39}{0.657/\sqrt{16}} = 1.735$. Since this value is not greater than 2.131, do not reject the null hypothesis. The data do not support the claim that the mean birth weight of male babies born to mothers taking a special vitamin supplement is different from 3.39 kg. The P-value can easily be obtained using software. For example, in Excel, in any cell enter the expression =TDIST(1.735,15,2) and press ENTER. The result is 0.1032. The numbers in the parentheses are the t statistic (entered as a positive number), the degrees of freedom, and a 2 to indicate a two-sided test.

23   Due to the size of the numbers in sample, this exercise should be done using software. For example, using Excel, enter the data in cells A1 to A6. In any cell, enter the expression =AVERAGE(A1:A6) and press ENTER. The result is the sample mean, 703.67. In another cell, enter the expression =STDEV(A1:A6) and press ENTER. The result is the sample standard deviation, 272.73.

H_0: $\mu = 1000$ hic; H_a: $\mu < 1000$ hic. The degrees of freedom is $n - 1 = 6 - 1 = 5$. The critical value from Table 10.1 is -2.015. $t = \dfrac{\bar{x}-\mu}{s/\sqrt{n}} = \dfrac{703.67-1000}{272.73/\sqrt{6}} = -2.661$. Since this value is less than -2.015, reject the null hypothesis. The data do support the claim that the mean hic measurement is less than 1000 hic. The P-value can easily be obtained using software. For example, in Excel, in any cell enter the expression =TDIST(2.661,5,1) and press ENTER. The result is 0.0224. The numbers in the parentheses are the t statistic (entered as a positive number), the degrees of freedom, and a 1 to indicate a one-sided test.

Section 10.2
Statistical Literacy and Critical Thinking

1   A two-way table contains entries that are frequency counts corresponding to two different variables. One variable is used for the row categories (Nausea and No Nausea) and the other variable is used for the column categories (Placebo and Chantix).

3 No. Both variables are observed and therefore we cannot conclude that there is a causal relationship between the two variables. The hypothesis test allows us to conclude that there is some relationship, but it does not allow us to conclude that there is a *causal* relationship.

5 The statement does not make sense. There is only one variable (gender) with two categories. There is no second variable that would be required for a two-way table.

7 This statement makes sense. If each observed value is close to the corresponding expected value, the terms $(O - E)^2/E$ will all be small and their sum χ^2 will also be small. This will result in failure to reject the null hypothesis of independence between the row and column variables.

Concepts and Applications

9 From Table 10.7, the critical value at the 0.05 significance level is $\chi^2 = 3.841$ for a 2 x 2 table. Since 3.499 is less than 3.841, do not reject the null hypothesis of independence. There is not sufficient evidence to claim that gender and response are somehow related.

11 From Table 10.7, the critical value at the 0.01 significance level is $\chi^2 = 6.635$ for a 2 x 2 table. Since 12.336 is greater than 6.635, reject the null hypothesis of independence. There is sufficient evidence to claim that gender and response are related.

13 a) H_0: The Polygraph indication and truth about lying are independent; H_a: The Polygraph indication and truth about lying are somehow related.

b) First we add a row and a column to the table to hold the totals.

	Did not Lie	Lied	Total
Polygraph says Lied	15	42	57
Polygraph says Not Lied	32	9	41
Total	47	51	98

Now P(Polygraph says Lied) = 57/98 = 0.5816 and P(Polygraph says Not Lied) = 41/98 = 0.4183, P(Not lied) = 47/98 = 0.4796, and P(Lied) = 51/98 = 0.5204.

Assuming that the null hypothesis is true that gender and student status are independent, P(Poly. Says Lie and Did not Lie) = P(Poly Says Lie) x P(Did not Lie) = (57/98)(47/98) and the expected value for Polygraph Say Lie and Did not Lie is (57/98)(47/98)(98) = 27.34. Similarly the expected values for the other three cells are

Poly. Says Lied and Lied = (57/98)(51/98)(98) = 29.66
Poly. Says Not Lied and Not Lied = (41/98)(47/98)(98) = 19.66
Poly. Says Not Lied and Lied = (41/98)(51/98)(98) = 21.34

Thus the table with observed and expected values is given by

	Did not Lie	Lied	Total
Polygraph says Lied	15(27.34)	42(29.66)	57
Polygraph says Not Lied	32(19.66)	9(21.34)	41
Total	47	51	98

Note that the column and row sums are the same for the observed and expected values.

c) To compute the χ^2 statistic, construct the following table, noting that it is a good idea to carry the expected values out to at least two decimal places:

Cell	O	E	O – E	$(O - E)^2$	$(O - E)^2/E$
Poly Lie/Did not Lie	15	27.34	-12.34	152.20	5.5674
Poly Lie/Lie	42	29.66	12.34	152.20	5.1308
Poly Not Lie/Did not Lie	32	19.66	12.34	152.20	7.7401
Poly Not Lie/Lie	9	21.34	-12.34	152.20	7.1330
Total	98	148.00	0.0		$25.5713 = \chi^2$

d)   At the 0.05 significance level for a 2 x 2 table, the critical value from Table 10.7 is 3.841.

e)   Since 25.5713 is greater than 3.841, reject the null hypothesis of independence. There is sufficient evidence to claim that the polygraph indication and the truth about lying are somehow related.

15 a) H_0: Response and job level are independent; H_a: Response and job level are somehow related.

b)   First we add a row and a column to the table to hold the totals.

	Yes	No	Total
Workers	192	244	436
Bosses	40	81	121
Total	232	325	557

Now P(Worker) = 436/557 and P(Boss) = 121/557
Also P(Yes) = 232/557 and P(No) = 325/557.
Assuming that the null hypothesis is true that response and job level are independent, P(Worker and Yes) = P(Worker) x P(Yes) = (436/557)(232/557) and the expected value for Workers and Yes is (436/557)(232/557)(557) = 181.60. Similarly the expected values for the other three cells are
Workers and No = (436/557)(325/557)(557) = 254.40
Bosses and Yes = (121/557)(232/557)(557) = 50.40
Bosses and No = (121/557)(325/557)(557) = 70.60
Thus the table with observed and expected values is given by

	Yes	No	Total
Workers	192(181.60)	244(254.40)	436
Bosses	40(50.40)	81(70.60)	121
Total	232	325	557

Note that the column and row sums are the same for the observed and expected values.

c)   To compute the χ^2 statistic, construct the following table, noting that it is a good idea to carry the expected values out to at least two decimal places:

Cell	O	E	O – E	$(O - E)^2$	$(O - E)^2/E$
Workers/Yes	192	181.6014	10.3986	108.1301	0.5954
Workers/No	244	254.3986	-10.3986	108.1301	0.4250
Bosses/Yes	40	50.39856	-10.3986	108.1301	2.1455
Bosses/No	81	70.60144	10.3986	108.1301	1.5316
Total	557	557.00	0.0		$4.698 = \chi^2$

d)   At the 0.05 significance level for a 2 x 2 table, the critical value from Table 10.7 is 3.841.

e)   Since 4.698 is greater than 3.841, reject the null hypothesis of independence. There is sufficient evidence to claim that response and job level are somehow related.

17   a) H_0: Improvement is independent of whether the participant was given the drug or the placebo; H_a: There is some relationship between improvement and treatment received.

b)   First we add a row and a column to the table to hold the totals.

	Improvement	No Improvement	Total
Drug	56	42	98
Placebo	49	43	92
Total	105	85	190

Now P(Drug) = 98/190 and P(Placebo) = 92/190
Also P(Improvement) = 105/190 and P(No Improvement) = 85/190.
Assuming that the null hypothesis is true that improvement and treatment are independent, P(Drug and Improvement) = P(Drug) x P(Improvement) = (98/190)(105/190) and the expected value for Drug and Improvement is (98/190)(105/190)(190) = 54.16. Similarly the expected values for the other three cells are
Drug and No Improvement = (98/190)(85/190)(190) = 43.84
Placebo and Improvement = (92/190)(105/190)(190) = 50.84
Placebo and No Improvement = (92/190)(85/190)(190) = 41.16
Thus the table with observed and expected values is given by

	Improvement	No Improvement	Total
Drug	56(54.16)	42(43.84)	98
Placebo	49(50.84)	43(41.16)	92
Total	105	85	190

Note that the column and row sums are the same for the observed and expected values.

c)   To compute the χ^2 statistic, construct the following table, noting that it is a good idea to carry the expected values out to at least two decimal places:

Cell	O	E	O − E	$(O − E)^2$	$(O − E)^2/E$
Drug/Improve	56	54.15789	1.84211	3.393352	0.062657
Drug/No Improve	42	43.84211	−1.84211	3.393352	0.077399
Placebo/Improve	49	50.84211	−1.84211	3.393352	0.066743
Placebo/No Improve	43	41.15789	1.84211	3.393352	0.082447
Total	190	190.00	0.0		0.289 = χ^2

d)   At the 0.05 significance level for a 2 x 2 table, the critical value from Table 10.7 is 3.841.

e)   Since 0.289 is less than 3.841, do not reject the null hypothesis of independence. There is not sufficient evidence to claim that there is a relationship between improvement and treatment received (drug or placebo).

19   a) H_0: The type of crime is independent of whether criminals are strangers; H_a: There is some relationship between the type of crime and whether criminals are strangers.

b)   First we add a row and a column to the table to hold the totals.

	Homicide	Robbery	Assault	Total
Stranger	12	379	727	1118
Acquaintance	39	106	642	787
Total	51	485	1369	1905

Now P(Stranger) = 1118/1905 and P(Acquaintance) = 787/1905
Also P(Homicide) = 51/1905, P(Robbery) = 485/1905, and P(Assault) = 1369/1905.
Assuming that the null hypothesis is true that the crime and whether the criminal is a stranger are independent, P(Stranger and Homicide) =

P(Stranger) x P(Homicide) = (1118/1905)(51/1905) and the expected value
for Stranger and Homicide is (1118/1905)(51/1905)(1905) = 29.93.
Similarly the expected values for the other five cells are
Stranger and Robbery = (1118/1905)(485/1905)(1905) = 284.64
Stranger and Assault = (1118/1905)(1369/1905)(1905) = 803.43
Acquaintance and Homicide = (787/1905)(51/1905)(1905) = 21.07
Acquaintance and Robbery = (787/1905)(485/1905)(1905) = 200.36
Acquaintance and Assault = (787/1905)(1369/1905)(1905) = 565.57
Thus the table with observed and expected values is given by

	Homicide	Robbery	Assault	Total
Stranger	12(29.93)	379(284.64)	727(803.43)	1118
Acquaintance	39(21.07)	106(200.36)	642(565.57)	787
Total	51	485	1369	1905

Note that the column and row sums are the same for the observed and
expected values.

c) To compute the χ^2 statistic, construct the following table, noting that
it is a good idea to carry the expected values out to at least two
decimal places:

Cell	O	E	O − E	$(O − E)^2$	$(O − E)^2/E$
Stranger/Homicide	12	29.9307	−17.9307	321.5103	10.74182
Stranger/Robbery	379	284.6352	94.3648	8904.721	31.28468
Stranger/Assault	727	803.4341	−76.4341	5842.175	7.27150
Acquaintance/Homicide	39	21.0693	17.9307	321.5103	15.25966
Acquaintance/Robbery	106	200.3648	−94.3648	8904.721	44.44254
Acquaintance/Assault	642	565.5659	76.4341	5842.175	10.32979
Total	1905	1905.00	0.0		119.330 = χ^2

d) At the 0.01 significance level for a 2 x 3 table, the critical value
from Table 10.7 is 9.210.

e) Since 119.330 is greater than 9.210, reject the null hypothesis of
independence. There is sufficient evidence to claim that there is a
relationship between crime category and whether criminals are
strangers.

Section 10.3
Statistical Literacy and Critical Thinking

1   ANOVA represents "analysis of variance." It is a method of testing the
equality of three or more population means from normal populations with equal
variances.

3   The null hypothesis is that the mean SAT scores for the three colleges are
equal; the alternative hypothesis is that the mean SAT scores are not all the
same at the three colleges.

5   This statement makes sense. A small P-value leads to rejection of the null
hypothesis of equal means.

7   This statement does not make sense. Analysis of variance can only tell us
whether there are differences in the mean weight loss from the four programs.
It cannot tell us that all four are effective.

Concepts and Applications

9   a) All of the population means of Flesch-Kincaid Grade Level scores are
equal for the three authors.

b) The population means of the Flesch-Kincaid Grade Level scores are not
all equal for the three authors.

c) The P-value is found at the right side of the table as 0.00077.

d)   Since the P-value is less than the 0.05 significance level, we reject the null hypothesis of equal population means for the Flesch-Kincaid Grade Level scores and conclude that the population means are not all equal for the three authors.

11 a)   The six colors of M&Ms have equal population mean weights.

b)   The six colors of M&Ms have population mean weights that are not all equal.

c)   The P-value is found near the right side of the table as 0.817.

d)   No. The very large P-value means that there is not sufficient evidence to conclude that the population mean weights for the six colors of M&Ms

13 We will use Excel to carry out the one-way ANOVA. Enter the names of the four categories of cars in Row 1 of columns A through D, and then enter the data values below the headings. Then click on Data (Tools for Excel 2003)and select Data Analysis from the drop-down menu. Click on Anova: Single Factor, and then click OK. In the box for Input Range, enter the expression A1:D6, and check the box for Labels in First Row. Now click in the circle for Output Range and enter F1 in the Output Range box. In the box for Alpha, enter the significance level 0.05 if it is not already there. Click OK.   The output is shown below.

Anova: Single Factor

SUMMARY

Groups	Count	Sum	Average	Variance
Subcompact	5	3344	668.8	58542.7
Compact	5	2779	555.8	8272.7
Midsize	5	2434	486.8	28110.2
Full-Size	5	2689	537.8	23905.2

ANOVA

Source of Variation	SS	df	MS	F	P-value	F crit
Between Groups	88425	3	29475	0.992167	0.42157	3.238867
Within Groups	475323.2	16	29707.7			
Total	563748.2	19				

The P-value is located in the lower right of the output and is given as 0.42157. Since this is larger than 0.05, do not reject the null hypothesis of equal means for the different weight categories. Based on these sample data, we cannot conclude that larger cars are safer.

15 We will use Excel to carry out the one-way ANOVA. Enter the names of the three categories of cars in Row 1 of columns A through C, and then enter the data values below the headings. Then click on Data (Tools in Excel 2003) and select Data Analysis from the drop-down menu. Click on Anova: Single Factor, and then click OK. In the box for Input Range, enter the expression A1:C10, and check the box for Labels in First Row. Now click in the circle for Output Range and enter F1 in the Output Range box. In the box for Alpha, enter the significance level 0.05 if it is not already there. Click OK.   The output is shown below.

Anova: Single Factor

SUMMARY

Groups	Count	Sum	Average	Variance
4000B.C.	9	1194	132.6667	17.5
1850 B.C.	9	1210	134.4444	11.27778
150 A.D.	9	1243	138.1111	22.61111

ANOVA

Source of Variation	SS	df	MS	F	P-value	F crit
Between Groups	138.7407	2	69.37037	4.04973	0.030518	3.402832
Within Groups	411.1111	24	17.12963			
Total	549.8519	26				

The P-value is located in the lower right of the output and is given as 0.030518. Since this is smaller than 0.05, reject the null hypothesis of equal means for the head breadths from the different epochs. Based on these sample data, there is sufficient evidence to conclude that there is a difference in the head breadth means from different epochs.

Chapter 10 Review Exercises

1   a)   There are 305 cell phone users, 23 of whom had an accident. Therefore 23/305 = 0.0754 or 7.54% of cell phone users had an accident. There are 453 who did not use a cell phone, 46 of whom had an accident. Therefore 46/453 = 0.1015 or 10.15% of non-users had an accident. Based on these data, cell phones do not appear to be dangerous. This assumes that the term "Cell phone user" means that those in this category used a cell phone while driving. There may be cell phone users who do not use them while driving.

b) H_0: Cell phone use is independent of whether people have car accidents; H_a: There is some relationship between cell phone use and having car accidents.

c)   P(Cell phone user) = 305/758
P(Not a cell phone user) = 453/758
P(Accident) = 69/758
P(No accident) = 689/798
If having an accident is independent of cell phone use, then the expected values are
Cell phone/Accident = (305/758)(69/758)(758) = 27.76
Cell phone/No accident = (305/758)(689/758)(758) = 277.24
No Cell phone/Accident = (453/758)(69/758)(758) = 41.24
No Cell phone/No Accident = (453/758)(689/758)(758) = 411.76

Cell	O	E	O − E	$(O - E)^2$	$(O - E)^2/E$
Cell/Accident	23	27.7639	−4.76385	22.69429	0.817404
Cell/No Accident	282	277.2361	4.76385	22.69429	0.081859
No Cell/Accident	46	41.2361	4.76385	22.69429	0.550349
No Cell/No Accident	407	411.7639	−4.76385	22.69429	0.055115
Total	758	758.00	0.0		$1.505 = \chi^2$

d) The value of the χ^2 statistic is 1.505.

e) From Table 10.7, we see that the χ^2 statistic is smaller than the critical value of 3.841 for a 0.05 significance level. Therefore the P-value is larger than 0.05.

f) Based on the preceding results, we cannot reject the null hypothesis that cell phone use and having an accident are independent. There is not sufficient evidence to conclude that there is some kind of relationship between cell phone use and having an accident in the past year.

g) More data suggest that cell phone usage while driving can be a distraction that may be dangerous. Some states have banned cell phone use by drivers.

3   The degrees of freedom is n − 1 = 8 − 1 = 7. From Table 10.1, the t value is 2.365. The margin of error is $= t\frac{s}{\sqrt{n}} = 2.365\frac{89.4793}{\sqrt{8}} = 74.818$. The 95% confidence interval is $\bar{x} \pm E = 77.375 \pm 74.818 = 2.557 \text{ to } 152.193$.

Chapter 10 Quiz

1   The population is normal, the sample size is small, and σ is unknown. Therefore the t distribution is the appropriate choice for a hypothesis test about a population mean.

3   Since the population is normal and the population standard deviation σ is known, the most appropriate distribution is the normal distribution for a hypothesis involving a claim about a population mean.

5   H_a: μ < 170

7   A one-way Analysis of Variance.

9   H_0: Sex and response are independent.